T0076066

Agile Visualization with Pharo

Crafting Interactive Visual Support Using Roassal

Alexandre Bergel

Apress®

Agile Visualization with Pharo: Crafting Interactive Visual Support Using Roassal

Alexandre Bergel
Santiago, Chile

ISBN-13 (pbk): 978-1-4842-7160-5
ISBN-13 (electronic): 978-1-4842-7161-2
https://doi.org/10.1007/978-1-4842-7161-2

Managing Director, Apress Media LLC: Welmoed Spahr
Acquisitions Editor: Steve Anglin
Development Editor: James Markham
Coordinating Editor: Mark Powers
Copyeditor: Kezia Endsley

Cover designed by eStudioCalamar

Cover image by Wirestock

Distributed to the book trade worldwide by Apress Media, LLC, 1 New York Plaza, New York, NY 10004, U.S.A. Phone 1-800-SPRINGER, fax (201) 348-4505, e-mail orders-ny@springer-sbm.com, or visit www.springeronline.com. Apress Media, LLC is a California LLC and the sole member (owner) is Springer Science + Business Media Finance Inc (SSBM Finance Inc). SSBM Finance Inc is a **Delaware** corporation.

For information on translations, please e-mail booktranslations@springernature.com; for reprint, paperback, or audio rights, please e-mail bookpermissions@springernature.com.

Apress titles may be purchased in bulk for academic, corporate, or promotional use. eBook versions and licenses are also available for most titles. For more information, reference our Print and eBook Bulk Sales web page at http://www.apress.com/bulk-sales.

Any source code or other supplementary material referenced by the author in this book is available to readers on GitHub. For more detailed information, please visit http://www.apress.com/source-code.

Printed on acid-free paper

Table of Contents

About the Author

Alexandre Bergel, Ph.D., is a professor (associate) at the Department of Computer Science (DCC), at the University of Chile and is a member of the Intelligent Software Construction laboratory (ISCLab). His research interests include software engineering, software performance, software visualization, programming environments, and machine learning. He is interested in improving the way we build and maintain software. His current hypotheses are validated using rigorous empirical methodologies. To make his research artifacts useful not only to stack papers, he co-founded Object Profile.

About the Technical Reviewer

Jason Whitehorn is an experienced entrepreneur and software developer and has helped many companies automate and enhance their business solutions through data synchronization, SaaS architecture, and machine learning. Jason obtained his Bachelor of Science in Computer Science from Arkansas State University, but he traces his passion for development back many years before then, having first taught himself to program BASIC on his family's computer while in middle school. When he's not mentoring and helping his team at work, writing, or pursuing one of his many side-projects, Jason enjoys spending time with his wife and four children and living in the Tulsa, Oklahoma region. More information about Jason can be found on his website: `https://jason.whitehorn.us`.

CHAPTER 1

Introduction

Computers are a formidable extension of the human brain: a computer liberates us from performing boring and repetitive tasks. Data visualization is a wonderful field where computers nicely complement what the brain excels at.

Conveying information through interactive visualizations is both a sophisticated engineering process and an art. When crafting a visualization, many decisions have to be made based on a carefully evaluated design aspect or a personal intuition. Either way, being able to quickly experiment with a new idea is key. Agile Visualization is about leveraging creativity by reducing the cost associated with building visualizations.

Visualizing data is probably the easiest part of the field of data visualization. Numerous books and sophisticated libraries exist for that very purpose. One of challenges of data visualization is to identify the right abstractions to build a visualization that is reusable, composable, extensible, navigable, and produced at a very low cost. Roassal is a visualization engine for Pharo and Smalltalk that leverages the experience of crafting and using data visualization.

Roassal is written in the Pharo programming language. All the examples provided in this book are therefore made for Roassal and are written in the Pharo programming language.

Since there is no better way than programming to craft a visualization, readers are expected to have some programming experience to fully enjoy Agile visualization. This book is written for a large audience, and it provides the necessary technical background as a starter for programming with Pharo.

Agile Visualization

Agile visualization promotes the creation of a visualization based on very short, incremental steps. A data analysis is carried out by building a number of visualizations, many of which are simply attempts and have a very short usage time. Reducing the creation time of a visualization to a few seconds or minutes greatly increases the number of different paths the data scientist can take to solve a given problem.

A. Bergel, *Agile Visualization with Pharo*, https://doi.org/10.1007/978-1-4842-7161-2_1

Jackson Pollock, a famous American painter, once said: "Splatter painting celebrates spontaneity, improvisation, and a highly physical approach to making art." Colors are thrown, mixed, and removed at will. This metaphor may be considered as a guiding line of Agile visualization. Similar to Agile programming, feedback should always occur a short time after the inception of a visualization.

The Pharo Programming Language

Roassal is developed in the Pharo programming language. All the source code provided in this book is written in Pharo. Pharo is a dynamically typed programming language, sharing some flavors with Smalltalk, Python, and Ruby. Pharo is easy to install, learn, and use. Pharo has a very simple syntax, which means that the code is understandable as soon as you have some programming knowledge. Pharo is both a programming language and a powerful environment. This book provides a light introduction to the syntax of Pharo and presents an overview on how to use and extend its environment.

If you do not know Pharo, here are a few pieces of advice. Pharo is easy to learn and use. It comes with fantastic programming tools to make you intimately interact with *objects*; an object being an elementary computational and logical unit. Resist the natural tendency to map your knowledge and expectations into Pharo. Embracing the way of thinking with objects is rich and enlightening. The Pharo programming environment is now your new friend, and plenty of great adventures will soon come.

The `https://pharo.org/download` website gives a very detailed instruction set to install Pharo. Installing Pharo is just a matter of a couple of clicks.

The Roassal Visualization Engine

Roassal is a visualization engine developed in Pharo. Roassal offers a simple API to build a visualization and maps arbitrary domain-specific objects to visual elements. This mapping process is at the core of Agile visualization and is extensively discussed in the book.

Roassal has a great list of features that make it share some similarities with other visualization engines, including D3.js and Matplotlib. However, Roassal naturally produces interactive visualizations to directly explore and operate on the represented domain objects. Furthermore, Roassal is integrated in the Pharo environment, which leverages the experience of building visualization.

At the moment this book is written, Roassal is the most commonly used visualization engine in Pharo and in the Smalltalk communities. Although alternatives are possible, including the Morphic framework or directly drawing in a `Form` object, Roassal offers a large set of pluggable and composable tools.

Roassal License

Roassal is distributed under the MIT License, which means that you can:

- Use Roassal for commercial purposes. Roassal can be freely distributed in a business-friendly manner, without any monetary payment due to the authors of Roassal.
- Modify the original source code of Roassal and distribute it as a separate project.

The MIT License is one of the most permissive software licenses available. However, the MIT License imposes two restrictions:

- You cannot hold the original authors of Roassal liable for any damage caused by using Roassal.
- You also cannot claim Roassal is your original work. Derivatives are okay as long as the original authors get credit and their name stays on the license.

Contributing to the Development of Roassal

Roassal is the result of more than ten years of hard work made by the authors of Roassal and the Pharo community. We encourage you to contribute to Roassal as well. It is easy to become a contributor of Roassal. There are many different ways to do so:

- If you find a bug or an opportunity for improvement, you can open an issue in the GitHub repository of Roassal that describes the issue.
- If you can improve the codebase of Roassal, define a pull request.

The GitHub repository of Roassal is available at `https://github.com/ObjectProfile/Roassal3`.

Accompanying Source Code

All the source code provided in this book is kept in the following Git repository: `https://github.com/bergel/AgileVisualizationAPressCode`.

It's better to use the provided material instead of trying to manually transcript the code given in the book.

Want to Have a Chat?

If you want to discuss Roassal, need help, or simply have a friendly chat, there are a number of ways to get in touch. Roassal has its own channel on the Pharo Discord server. The Pharo community extensively uses Discord to communicate. You can join the Pharo Discord server by following the instruction provided at `https://pharo.org/community`. After joining it, you can jump on to the channel `#roassal` and say "Hi!". A number of friends' channels are also in the Pharo Discord server:

- `#roassal-scriptoftheday` regularly provides short script that's ready to be executed. These scripts typically illustrate a particular aspect of Roassal.
- `#roassal-commit` reflects the activity of the Roassal GitHub repository by listing the commits made in the repository.

Adding the hashtag `#Roassal` and the `@Roassal1` users to your tweets is also a great way to advertise ideas. You can also use the mention `@PharoProject` since the Pharo community will surely have an interest in your post. Finally, you can reach me by email at `alexandre.bergel@me.com`.

Book Overview

Agile Visualization is divided into 15 chapters, each targeting a specific topic in the field of visualizing and interacting with data in Pharo:

- Chapter 2 provides a tour of Roassal by presenting a few visualizations.
- Chapter 3 is an introduction to the Pharo programming language.

- Chapter 4 discusses the notion of Agile visualization.
- Chapter 5 presents the relevant components of Roassal.
- Chapter 6 explains the Roassal Canvas.
- Chapter 7 lists the shapes offered by Roassal.
- Chapter 8 describes the line builder.
- Chapter 9 details the composition mechanism of shapes.
- Chapter 10 presents the normalizers and scales offered by Roassal.
- Chapter 11 highlights the most relevant interactions supported by Roassal.
- Chapter 12 lists a number of commonly used layouts.
- Chapter 13 describes the integration of Roassal in the Inspector framework of Pharo.
- Chapter 14 applies visualizations to explain the behavior of reinforcement learning, a machine learning algorithm.
- Chapter 15 details how visualizations can be automatically generated from a GitHub repository using GitHub Actions.

As with any successful open source project, Roassal is driven by active community effort. The positive aspect of being a successful open source project is that Roassal is evolving every day (literally). The negative aspect is that documentation quickly becomes obsolete. The book is written in a way that deep technical aspects are not discussed while general concepts are largely presented. These general concepts are much more stable over time.

Who Should Read This Book?

This book is designed to satisfy various facets of a large audience:

- *Data scientists* -Readers familiar with Pharo will learn the essential components of the Roassal visualization engine. After reading the book, you will be able to apply visualization techniques to any domain data. Many examples are provided with the book and they can be used as a guide or as templates for analyses.

- *Designers of visualization engine* – Designing and implementing a visualization engine is an incredibly difficult task. Roassal went through three complete rewrites to reach its current status. Readers who want to implement an engine may definitely find valuable resources regarding the design and implementation of a visualization engine.

Acknowledgments

The book is the result of multiple and long-lasting collaborations. First of all, I would like to thank the Lam Research company. Lam Research's team has always been supportive, both psychologically and financially. Thanks CH and Chris! You made this book a reality.

Many people in the Moose, Pharo, and ESUG communities have deeply contributed to what is presented in the book. Your enthusiastic support and trust in what I do has always been invaluable.

I also would like to thank you, yes you, the reader, for your questions, support, bug fixes, contribution, and encouragement.

I am also deeply grateful to the following people for their contributions (in no particular order): CH Huang, Chris Thorgrimsson, Milton Mamani, Ted Brunzie, Tudor Gîrba, Renato Cerro, Stéphane Ducasse, Yuriy Tymchuk, Natalia Tymchuk, Juraj Kubelka, Juan Pablo Sandoval Alcocer, Vanessa Peña, Ronie Saldago, Alvaro Jose Peralta, Pablo Estefo, Igor Stasenko, Faviola Molina, Ricardo Jacas, Daniel Aviv Notario, Sergio Maass, Serge Stinckwich, Bui Thi Mai Anh, Nick Papoulias, Johan Fabry, Nicolai Hess, Miguel Campusano, Peter Uhnák, Martin Dias, Jan Blizničenko, Samir Saleh, Nicolai Hess, Leonel Merino, Volkert, Pierre Chanson, Andrei Chis, Thomas Brodt, Mathieu Dehouck, Miguel Campusano, Onil Goubier, Thierry Goubier, Esteban Maringolo, Alejandro Infante, Philippe Back, Stefan Reichhart, Ronie Salgado, Steffen Märcker, Nour Jihene Agouf, Pavel Krivanek, and Esteban Lorenzano.

CHAPTER 2

Quick Start

This chapter provides an overview of Roassal, as well as a few examples. The chapter does not go into detail since that is something that will occur in forthcoming chapters. Many short code snippets are provided and briefly described. You can copy and paste these snippets directly into Pharo and each one illustrates a particular aspect of the Roassal platform.

It is important to keep in mind that this chapter is just a tour of Roassal and Pharo. If you are not familiar with Pharo, you might find the amount of code given here a bit confusing. The next chapter serves as an introduction of the Pharo language and explains many aspects of the language syntax.

All the code provided in this chapter is available at `https://github.com/bergel/AgileVisualizationAPressCode/blob/main/01-02-QuickStart.txt`.

Installation

Roassal is the framework for the Pharo programming language. As such, the first step to try the examples given in this chapter is to install Pharo. The Pharo website provides all the necessary instructions to do so (`https://pharo.org/download`). Pharo can be installed directly via the command line or using the Pharo launcher. Both ways are equivalent and you may prefer one over the other based on your personal workflow.

Once you have installed Pharo, you need to open the Playground, which is a tool offered by Pharo to execute code. You can think of the Playground as a UNIX terminal. The Playground is opened from the top toolbar, as shown in Figure 2-1.

A. Bergel, *Agile Visualization with Pharo*, https://doi.org/10.1007/978-1-4842-7161-2_2

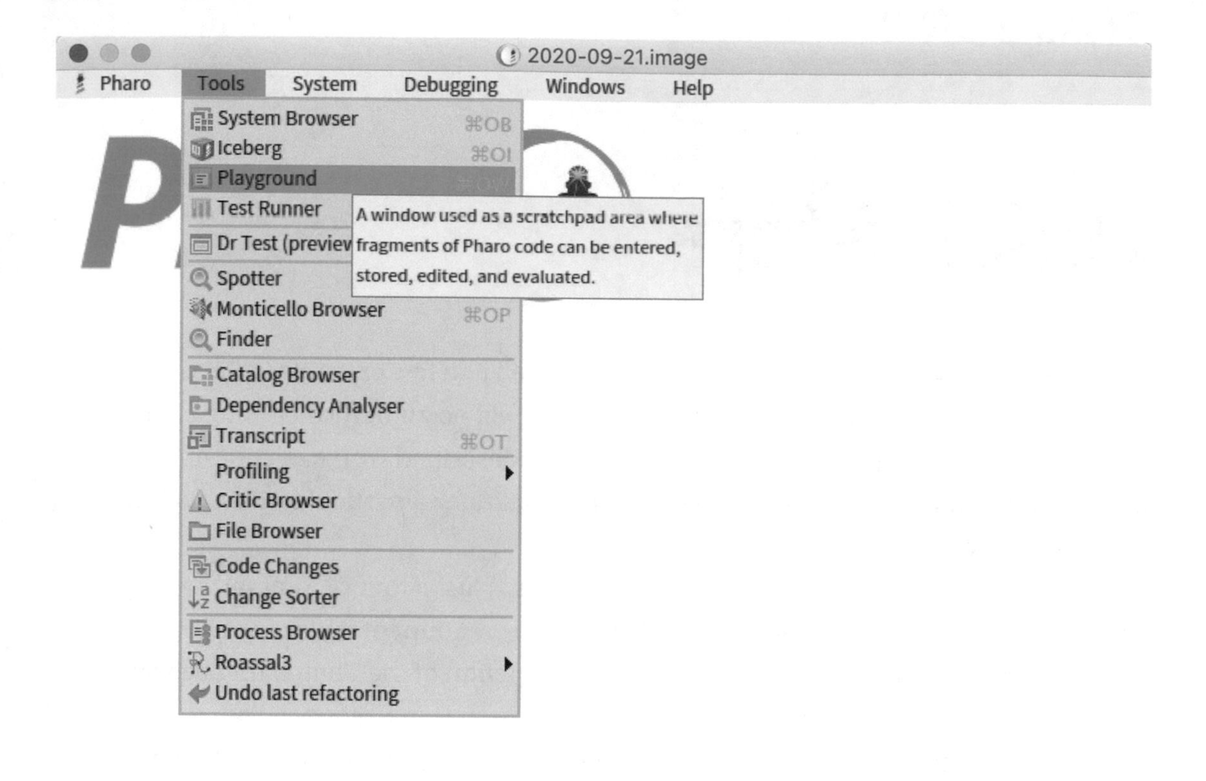

Figure 2-1. *Opening the Playground from the World menu*

After selecting Playground from the menu, you'll see a window in which you can type any Pharo instructions. Just type the following instructions (or copy and paste them if you are reading an electronic version of this book):

```
[ Metacello new
    baseline: 'Roassal3';
    repository: 'github://ObjectProfile/Roassal3';
    load: 'Full' ] on: MCMergeOrLoadWarning do: [ :warning | warning load ]
```

You can execute the code by pressing Cmd+D (if you are a macOS user) or Ctlr+D (for Windows and UNIX users), or by pressing the green Do It button.

Loading Roassal should take a few seconds, depending on your Internet connection. Once it's loaded, you may want to save your Pharo environment (a.k.a , the image in the Pharo jargon) by choosing Pharo➤Save from the toolbar menu. You are now ready to try your first visualization.

First Visualization

Most of the visualizations in this book are written as short scripts, directly executable in the Playground. You can open a new Playground or simply reuse an already open Playground (e.g., the one you used to load Roassal). In that case, you should erase the contents of the Playground and paste in the new script.

Evaluate the following example in the Playground (see Figure 2-2).

```
c := RSCanvas new.
1 to: 100 do: [ :i |
    c add: (RSLabel new model: i) ].

RSLineBuilder line
    shapes: c nodes;
    connectFrom: [ :i | i // 2 ].

RSClusterLayout on: c nodes.
c @ RSCanvasController.
c open
```

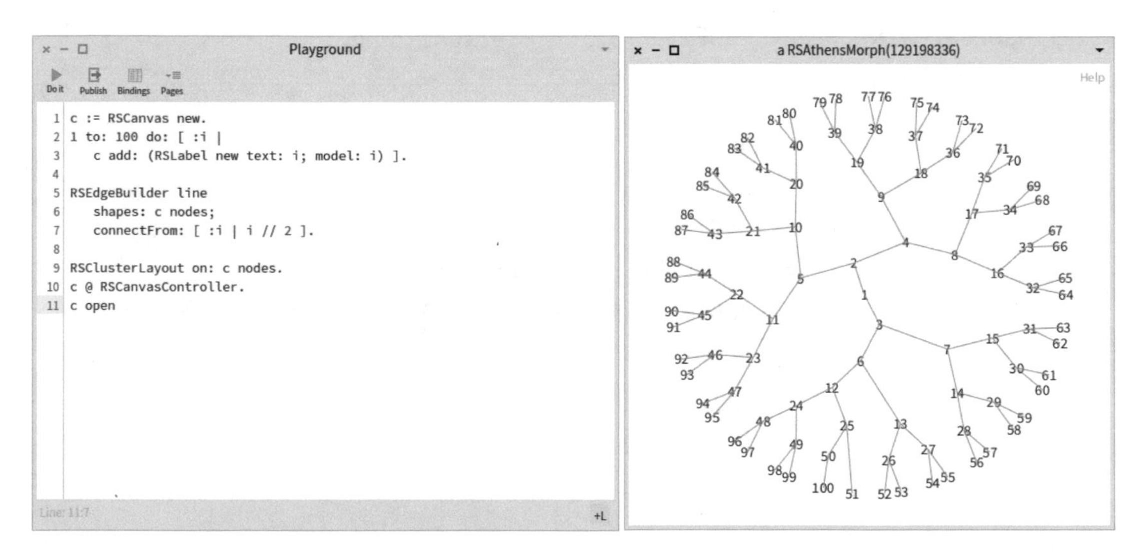

Figure 2-2. *Connecting numbers*

The script begins by creating a canvas using the RSCanvas class. The script adds 100 labels to the canvas, each representing a number between 1 and 100. Lines are built as follows: for each number i between 1 and 100, a line is created from the element representing i //2 and i. The expression a //b returns the quotient between a and b, e.g., 9 //4 = 2 and 3 //2 = 1. Nodes are then ordered using a cluster layout.

As a short exercise, you can replace 100 with any other value. You can also replace the RSClusterLayout class with RSRadialTreeLayout, RSTreeLayout, or RSForceBasedLayout.

Visualizing the Filesystem

You will reuse the previous visualization to visualize a filesystem instead of arbitrary numbers. Pharo offers a complete library to manipulate files and folders. Integrating files into a Roassal visualization is easy. Consider the following script:

```
path := '/Users/alexandrebergel/Desktop'.
extensions :=
    { 'pdf' -> Color red . 'mov' -> Color blue } asDictionary.
allFilesUnderPath := path asFileReference allChildren.
c := RSCanvas new.

allFilesUnderPath do: [ :aFile |
    | s color |
    s := RSEllipse model: aFile.
    color := extensions at: aFile path extension
                        ifAbsent: [ Color gray ].
    s color: color translucent.
    s @ RSPopup @ RSDraggable.
    c add: s ].

RSNormalizer size
    shapes: c nodes;
    from: 10; to: 30;
    normalize: [ :aFile | aFile size sqrt ].
```

```
RSLineBuilder line
    shapes: c nodes;
    connectFrom: #parent.

RSClusterLayout on: c nodes.
c @ RSCanvasController.
c open
```

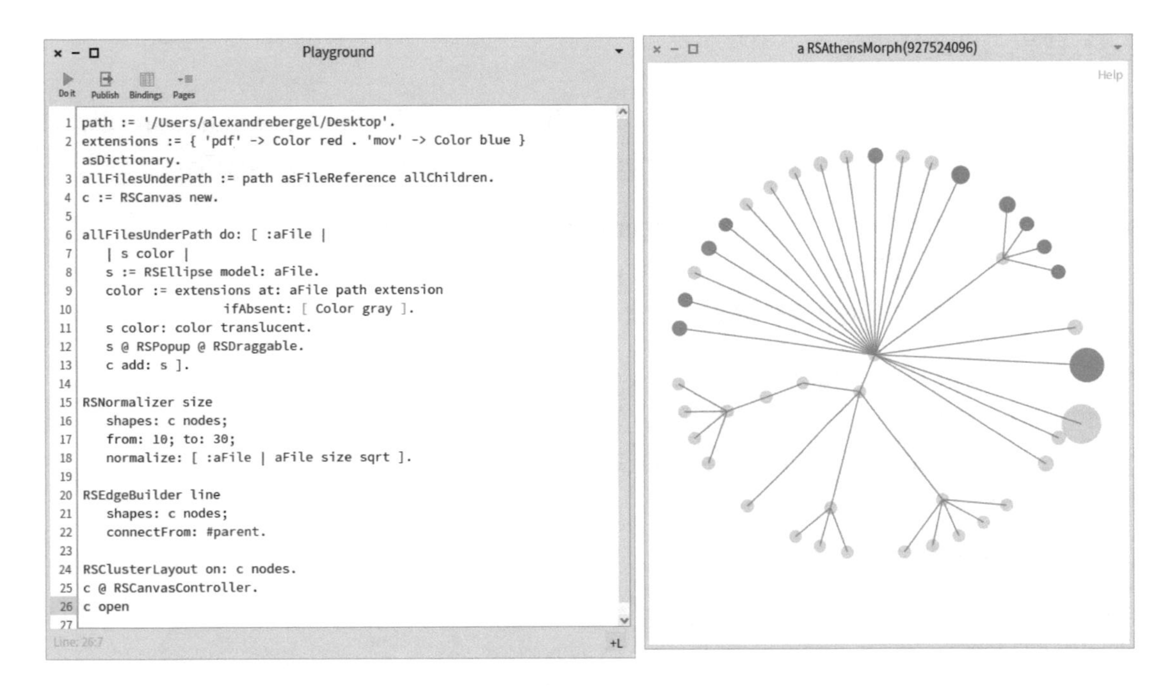

Figure 2-3. *Visualizing the filesystem*

Figure 2-3 shows the contents of the path `/Users/alexandrebergel/Desktop`, which correspond to the contents of the desktop on macOS. The `path` variable contains a location on your filesystem. Obviously, you need to change the path to execute the script. Note that indicating a large portion of the filesystem may significantly increase the computation time since recursively fetching file information is time-consuming. The `path asFileReference` expression converts a string indicating a path as a file reference. `FileReference` is a Pharo class that represents a file reference, typically locally stored on hard disk. The `allChildren` message gets all the files recursively contained in the path. The visualization paints files whose names end with `.pdf` in red and paints all videos files ending with `.mov` in blue.

Compared to the previous example, this visualization uses a normalizer to give each circle a size according to the file size. The size varies from 10 to 30 pixels, and it uses a square root (sqrt) transformation to cope with disparate sizes.

As an exercise, you can extend the color schema for specific files located on your filesystem. Color rules must follow the pattern `'pdf'-> Color red` and must be separated by a period character.

Charting Data

Roassal offers a sophisticated library to build charts. Consider the following example, showing the high of the COVID-19 pandemic during its first 250 days (see Figure 2-4).

```
url := 'https://raw.githubusercontent.com/ObjectProfile/',
       'Roassal3Documentation/master/data/',
       'covidDataUntil23-September-2020.txt'.

rawData := OpalCompiler evaluate: ((ZnEasy get: url) contents).
countries := rawData collect: #first.
allData := rawData collect: #allButFirst.
color := NSScale category20.

chart := RSChart new.
chart extent: 400 @ 400.
chart colors: color.
allData do: [ :data | chart addPlot:(RSLinePlot new y: data) ].
chart xlabel: 'Days since epoch' offset: 0 @ 20.
chart ylabel: 'Contaminated' offset: -60 @ 0.
chart title: 'Coronavirus confirmed cases'.
chart addDecoration: (RSHorizontalTick new fontSize: 10).
chart addDecoration: (RSVerticalTick new integerWithCommas; fontSize: 10).
chart ySqrt.
chart build.

b := RSLegend new.
b container: chart canvas.
countries with: chart plots do: [ :c : p |
    b text: c withBoxColor: (chart colorFor: p) ].
```

```
b layout horizontal gapSize: 30.
b build.
b canvas open
```

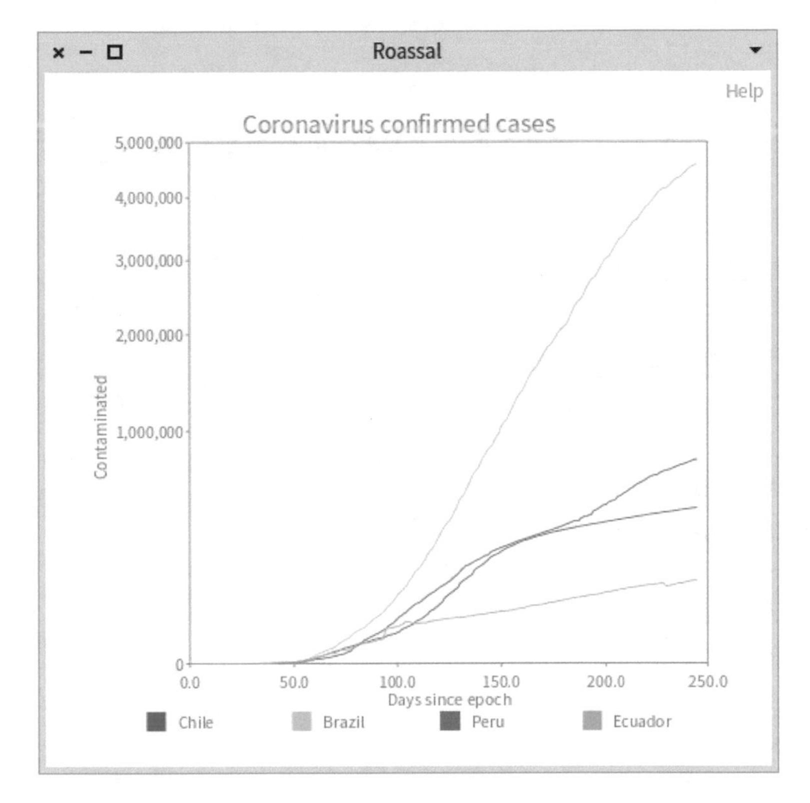

Figure 2-4. *The first 250 days of the pandemic in 2020*

The `url` variable points to a dataset that contains the first 250 days of the COVID-19 pandemic in 2020 caused by the SARS-CoV-2. Note that we split the URL to accommodate the book formatting. The data stored in the `covidDataUntil23-September-2020.txt` file is a simple serialization of the data to be rendered by the charter. The `RSChart` class is the entry point of the charting library. Plots are added to a chart and a few decorations are added. A legend is located below to associate curves with countries.

Sunburst

A sunburst is a visualization designed to represent hierarchical data structure. Consider the following example (Figure 2-5).

```
sb := RSSunburstBuilder new.
sb sliceShape withBorder.
sb sliceColor: [ :shape | shape model subclasses isEmpty
                    ifTrue: [ Color purple ]
                    ifFalse: [ Color lightGray ] ].
sb explore: Collection using: #subclasses.
sb build.
sb canvas @ RSCanvasController.
RSLineBuilder sunburstBezier
    width: 2;
    color: Color black;
    markerEnd: (RSEllipse new
        size: 10;
        color: Color white;
        withBorder;
        yourself);
    canvas: sb canvas;
    connectFrom: #superclass.
sb canvas open
```

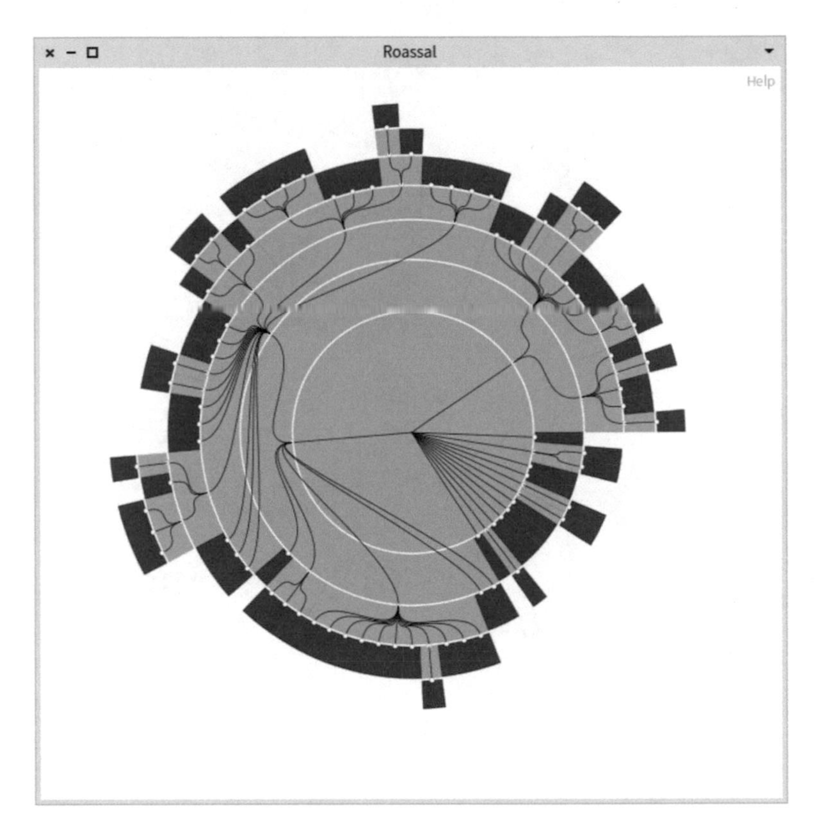

***Figure 2-5.** Visualizing the collection class hierarchy using a sunburst*

This sunburst is a software visualization. Each arc represents a class, and the nesting indicates class inheritance, which is highlighted with Bezier lines.

Graph Rendering

Roassal offers a wide range of tools to manipulate and render graphs. Consider the following script (see Figure 2-6).

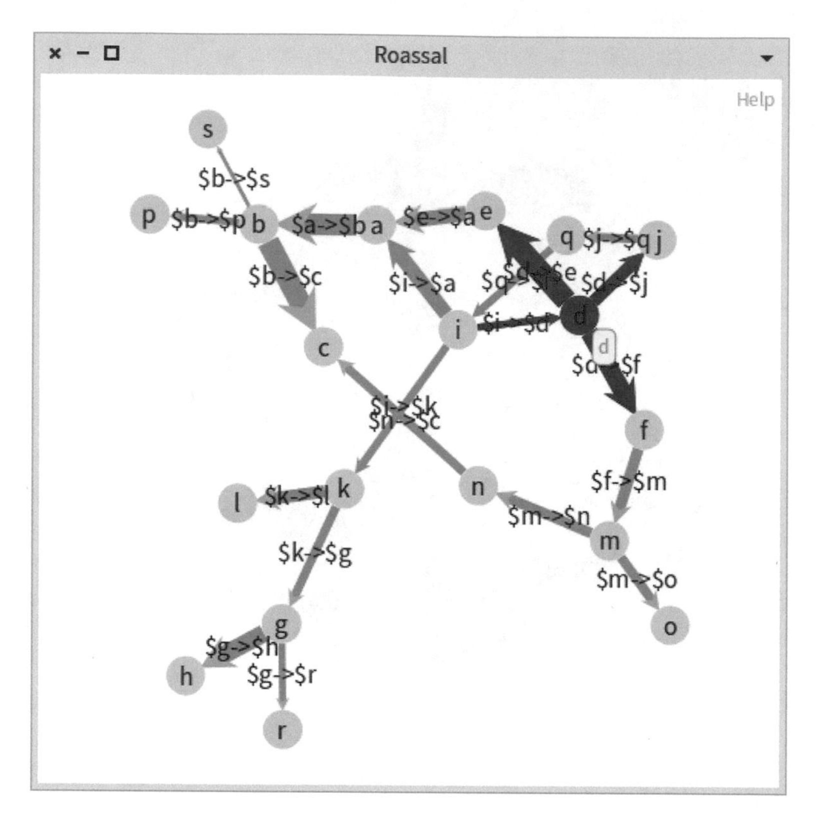

Figure 2-6. *Visualizing a graph*

```
nodesModel := $a to: $s.
edges := #( #( $a $b 30 ) #( $b $s 1 ) #( $b $p 4 ) #( $b $c 30 )
            #( $d $e 30 ) #( $d $f 20 ) #( $d $j 10 ) #( $e $a 15 )
            #( $f $m 8 ) #( $g $h 20 ) #( $g $r 3 ) #( $i $a 14 )
            #( $i $k 4 ) #( $i $d 3 ) #( $j $q 5 ) #( $k $l 10 )
            #( $k $g 5 ) #( $m $n 7 ) #( $m $o 6 ) #( $n $c 5 )
            #( $p $b 5 ) #( $q $i 4 ) ).

graph := Dictionary new.

nodesModel do: [ :aNode |
    graph at: aNode put: Set new ].

edges do: [ :edge |
    fromNode := edge first.
    toNode := edge second.
    (graph at: fromNode) add: toNode ].
```

```
canvas := RSCanvas new.
nodes := RSCircle models: (nodesModel) forEach: [:circle :model | circle
size: 20; color: Color veryLightGray. ].

nodes @ RSDraggable; @ RSPopup.
canvas addAll: nodes.

highlightable := RSHighlightable new.
highlightable highlightColor: Color red.
highlightable withEdges.
nodes @ highlightable.

lb := RSLineBuilder line.
lb canvas: canvas.
lb makeBidirectional.
lb moveBehind.
lb objects: nodesModel.
lb connectToAll: [ :aNumber | graph at: aNumber ].
canvas lines do: [ :line | | edge length |
    edge := edges detect: [ :e |
        e first = line model key
            and: [ e second = line model value ] ].
    length := edge third sqrt * 2.
    line width: length.
    line attachPoint: (RSBorderAttachPoint new
        endOffset: length).
    line markerEnd: (RSShapeFactory arrow size: length * 2).
    line markerEnd offset: length / -5.
     ].
(canvas nodes, canvas lines) @ (RSLabeled new
    in: [ :lbl |
        lbl location middle.
        lbl shapeBuilder labelShape color: Color black ];
    yourself).
RSForceBasedLayout new charge: -500; doNotUseProgressBar; on: nodes.
canvas @ RSCanvasController.
canvas open
```

Nodes considered in the graph represent characters ranging from `$a` to `$s`. The edges variable defines the weighted connections between nodes. The script renders the graph using labeled arrow lines and uses a force-based layout. Furthermore, moving the mouse above a node highlights lines connected to it.

What Have You Learned in This Chapter?

Many examples are available in the Roassal distribution, in the `Roassal3-Examples` package. Hundreds of examples cover many different parts of Roassal. Many relevant topics in Roassal are illustrated by more than one example to illustrate its flexibility and the capability to configure.

Some of the examples in this chapter are covered in-depth in coming chapters. I recommend you experiment by adapting and tweaking these examples.

CHAPTER 3

Pharo in a Nutshell

Programming is the skill that allows you to tell a computer what it has to do. Effective data visualization cannot be properly exercised without programming. This chapter is an introduction to the Pharo programming language. Although I have tried to make this chapter smooth and easy to read, having some basic programming knowledge will help you better understand it, and indeed the book as a whole.

Pharo is an object-oriented, class-based, dynamically typed programming language. Pharo promotes live programming at its heart, and it does this very well. If these statements do not make much sense to you, no worry, as you will explore these important topics in this chapter.

The chapter begins with the ubiquitous "hello world" example accompanied by a brief introduction of what programming with objects is all about. The focus of the chapter is about Pharo itself; you will learn how to develop a small application to visualize tweets.

All the code provided in this chapter is available at `https://github.com/bergel/AgileVisualizationAPressCode/blob/main/01-03-PharoInANutshell.txt`.

Hello World

Instead of giving a long rhetorical description of object-orientation, this section uses a simple example. You need to write code in the Playground. You can open the Playground from the menu toolbar. The code is executed by pressing the green button (Do It) or by pressing Cmd+G or Control+G, depending on whether you use macOS or Windows/Linux, respectively. Type and execute a very simple Pharo script, as shown in Figure 3-1.

```
c := RSCanvas new.
c add: (RSLabel new fontSize: 30; text: 'Hello World').
c openOnce
```


A. Bergel, *Agile Visualization with Pharo*, https://doi.org/10.1007/978-1-4842-7161-2_3

Executing the script should open a new window in which you see the text `Hello World`, as shown in Figure 3-1.

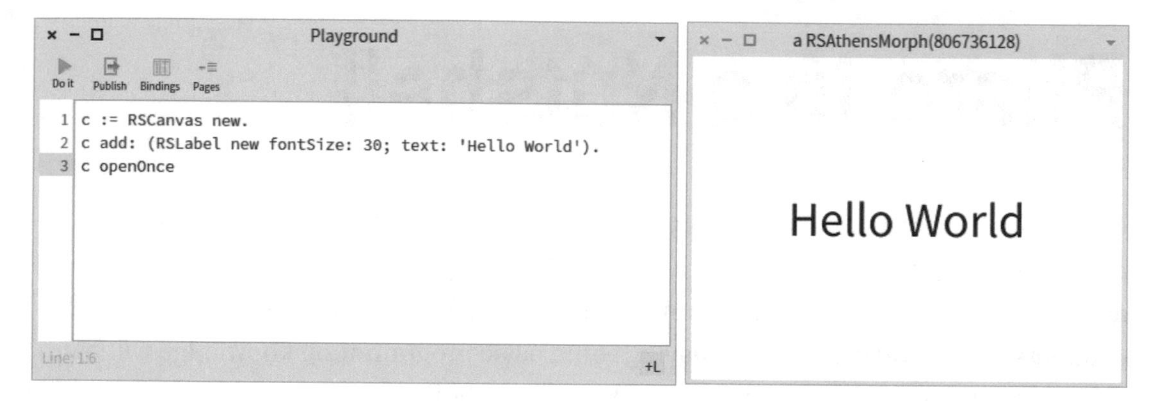

Figure 3-1. *Hello World*

`RSCanvas` refers to a class. A class is an object factory, and creating an object is like baking cakes. All the cakes produced in a pan have the same physical aspects, but attributes, such as flavors and colors, may vary. A class can produce objects that are different in the values of the variables; however, the variable names and the number of variables remain constant among all objects issued from the same class. A class is easily recognizable in Pharo code because its name always begins with a capital letter. The code given previously uses two classes, `RSCanvas` and `RSLabel`.

The first line creates a canvas object. An object is created by sending the `new` message to a class. For example, the `RSCanvas new` expression sends the `new` message to the `RSCanvas` class (which is also an object by the way). As such, it creates a canvas. Similarly, `String new` creates an empty string character and `Color new` creates a black color. The canvas, from executing `RSCanvas new`, is said to be the object produced by `RSCanvas`.

Objects interact with each other by sending messages. Consider the expression `'hello world' asCamelCase`, which evaluates to `'HelloWorld'` (simply type `'hello world'asCamelCase` in the Playground, right-click the line and execute the item `Print it`). This expression sends the `asCamelCase` message to the string object `'hello world'`. In Pharo, a class is also an object, which means that objects are created by sending a message to a class. The `new` message is sent to the `RSCanvas` class and it has the effect of creating an object canvas. This object is assigned to the variable `c` using the assignment operator `:=`.

In the second line of this script, the new message is sent to the RSLabel class, which creates an object called label. The fontSize: message is sent to that object with 30 as its argument, which has the effect of setting the font size to 30 units. Thanks to the cascade operator (the semicolon ;), the Hello World message text is sent to the label. It configures the label with the provided string. The configured label is then added to the c canvas when it's provided to the add: message.

Visualizing Some Numbers

Consider this second example, which has been slightly more elaborated (see Figure 3-2).

```
c := RSCanvas new.
#(20 40 35 42 10 25) do: [ :v |
    c add: (RSCircle new size: v; model: v).
].
RSHorizontalLineLayout on: c shapes.
c shapes @ RSLabeled.
c @ RSCanvasController.
c openOnce
```

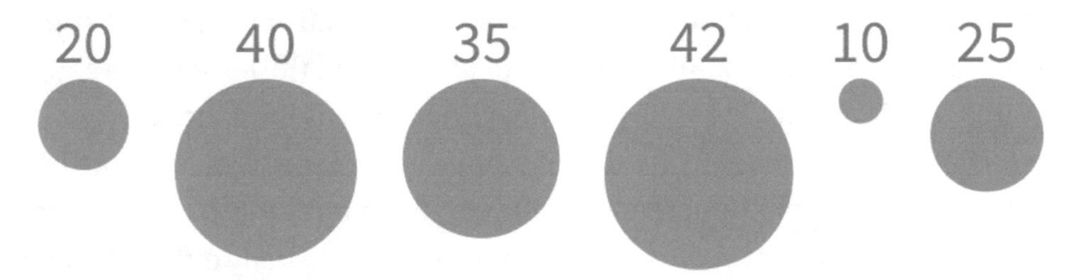

Figure 3-2. *Circles representing numbers*

This example renders six circles, each having a particular size. The expression #(20 40 35 42 10 25) defines an array containing some numbers. A collection can be iterated using the do: message with a one-parameter block as its argument. This block is considered a function that is executed for each value contained in the array. An expression such as [:v | ...] defines such a block that accepts one argument. A block accepting two arguments would be written as [:arg1 :arg2 | ...] in which the arguments are arg1 and arg2.

The RSCircle new size: v; model: v expression creates a circle with a determined size and model object. In Roassal, each visual element can represent an object, which is designed as a model. The model is used when applying interactions, such as RSLabeled.

Roassal offers several layouts to arrange shapes in a two-dimensional space. This example applies the horizontal line layout to all the shapes contained in the canvas. The c shapes @ RSLabeled expression sends the shapes message to the canvas c to obtain all the shapes contained in the canvas. The @ RSLabeled message is sent to the collection of shapes, which individually add the RSLabeled interaction to each shape. This interaction locates a textual label on each shape. The name of the message is @, while the argument is the RSLabeled class.

From Scripts to Object-Oriented Programming

We informally defined a set of instructions ready to be typed and evaluated in the Playground as a *script*. Scripts are great because they are usually self-contained and contain all the necessary logic to be easily understood.

Most visualization engines and data analysis environments operate on this principle: scripts are typed in a workspace or a web page and are then executed to produce a visualization. This approach to build a visualization or a small application is appealing since you do not need to look somewhere else to understand the logic of the script. However, this way of developing software artifacts has serious limitations.

Long scripts are difficult to maintain and modify in the long term. Imagine coming back to a script that's several hundred lines of code after a few months have passed. Understanding the reasons and the structure of the script is usually difficult. So many details are provided in a small portion of the screen! If it's not properly structured, adapting a complex visualization may consume a ridiculously large amount of time. This situation is well known to data scientists and software engineers. Fortunately, a couple of decades ago, the software engineering research community produced a way of programming that an cope with the inherent complexity of software artifact development. Object-oriented programming is the most successful way to handle complex and large development. As such, I encourage you to apply object-orientation to build visualization instead of using linear scripts.

Pillars of Object-Oriented Programming

Object-oriented programming simplifies the programming activity. Handling objects, instead of functions or code snippets, involves a metaphor that is familiar to humans: an object may act on some actions, have a behavior of its own, and hide details about how it is physically built.

Let's bring a bit of theory into this. There are five essential ingredients of an object-oriented system:

- *Encapsulation*: In your daily life, you are used to handling information that is not publicly accessible, e.g., your social security number, bank account number, and so on. Encapsulation in object-oriented systems is about letting objects have private information. Private information may reflect detail that is not directly necessary to a service consumer. When private information has to be publicly exposed, asking a question is the polite and cordial way to access it. In object-oriented programming, sending a message is the way to obtain information and carry out any computation. Encapsulation in object-oriented programming enables abstractions and information hiding, considerably easing the maintenance and evolution of software artifacts.
- *Composition*: A complex problem may be solved by cutting it into smaller problems. Once these smaller parts are solved, a number of independent results have to be composed to form the result of a complex problem.
- *Distribution of responsibilities:* In your daily life, you have duties and responsibilities. Having a clear separation of concerns is key to having a good object-oriented design, which also greatly contributes to understanding and maintaining software artifacts. For example, instead of asking someone's weight in order to select what may be eaten, it is better for everybody to let that person make a responsible choice. This example is not far stretched: many difficulties in software maintenance are directly rooted from improperly assigned responsibilities in software.

- *Message sending:* Electronic emails are the base of daily communication. A person, called the sender, sends an email to another person, called the receiver. In object-oriented programming, objects communicate in a similar fashion. Computation is carried out by sending messages between objects. An object sends messages to other objects. After sending a message, a reply is returned. In object-orientation, sending a message is often perceived as a way to delegate responsibilities.
- *Inheritance*: This is a general concept that is specialized to address particular requirements. Inheritance allows you to define a conceptual hierarchy, reuse code, and support polymorphism. Inheritance may say that an ellipse and text are two graphical shapes.

These five pillars are not tied to one programming language. So, in theory, it is perfectly okay to have an object-oriented design in a procedural language such as C. However, having a programming language that enforces these principles greatly alleviates the programmer's task.

There are numerous object-oriented languages around and Pharo is one of them. Pharo differs from other languages by offering an homogeneous way of expressing computation: everything is an object, therefore computation only happens by sending messages. When objects are taken seriously, there is no need for primitive types (e.g., `int` and `float`) or language operators! Having to deal only with message sending significantly reduces the amount of technological details associated with the program execution.

The following sections explain some of these concepts and illustrate how Pharo supports an elegant way to program with objects.

Sending Messages

Sending a message is the elementary unit of computation. Understanding how to send a message is key to feeling comfortable in Pharo. Consider the expression:

```
'the quick brown fox jumped over the lazy dog' copyReplaceAll: 'fox'
with: 'cat'
```

This expression sends a message to the string object `'the quick brown fox jumped over the lazy dog'`. The message has the name `#copyReplaceAll:with:` and two arguments, `'fox'` and `'cat'`, which are two string objects. The result of sending this message is `'the quick brown cat jumped over the lazy dog'`, another string.

In Pharo, a character string (often simply called a *string*) is a collection of characters written between two quote marks (e.g., `'fox'`). A string is an object, which means you can send messages to it. In Pharo, a message is composed of two essential ingredients: a name and a collection of arguments. It may be that the set of arguments is empty. For example, the `'fox'asUppercase` expression, which evaluates to `'FOX'`, sends the `#asUppercase` message to the `'fox'` string and no arguments are involved here.

Message sending is at the heart of the Pharo programming language, and as such, messages are well expressed in its syntax. There are three kinds of messages that can be sent:

- *Unary messages:* A unary message does not take an argument. The expression `'fox'asUppercase` sends a unary message to the string `'fox'`.
- *Binary messages:* A binary message has exactly one argument and its name is not made of alphanumerical characters. Instead, one or two characters are common for binary messages, such as `+`, `/`, `-`, `<`, and `>>`. The expression `2 + 3` sends a binary message named `+` with the argument `3` to the object `2`. You may notice that this expression is therefore semantically different from `3 + 2`, although the result is obviously the same. Note that the expression `3 + 2 * 2` returns `10`, and not `7` as you may expect. If you want to enforce mathematical priorities in arithmetic operations, use parentheses, as in `3 + (2 * 2)`. In practice, not having priorities between mathematical operators is not a problem. Inserting parentheses is cheap and does not hamper readability (most of the time it increases readability).
- *Keyword messages:* A keyword message is neither unary nor binary. A keyword message accepts one or more object arguments. Consider the example `'the quick brown fox jumped over the lazy dog'includesSubstring 'fox'`. This expression evaluates to true.

The name of the keyword message is #includesSubstring: and the argument is 'fox'. Each argument is preceded by a keyword. For example, the message replaceAllRegex : 'fox'with: 'cat' contains two keywords and therefore two arguments. Arguments are inserted in the message name.

Sending a message triggers a mechanism that searches for the adequate method to execute. This mechanism, often called *method lookup,* begins at the class of the object and follows the superclass link. If a method with the same name as the message is not found in the class of the object, the method lookup searches in the superclass. The process is repeated until the method is found. When a message sent, the keyword super triggers a lookup that begins in the superclass of the class that contains the call on super.

Creating Objects

An object is a bundle of data to which messages can be sent. Most of the time, an object is created by simply sending the new message to a class. This reveals the true nature of classes, being an object factory. Objects produced from a unique class are different but understand the same set of messages and have the same variables. Differences between two or more objects issued from the same class are the values given to these variables. For example, consider the following expression:

```
Object new == Object new
```

This expression sends three messages, the new message twice and the == message once, to compare object identities. The expression evaluates to false, since the two objects are different, i.e., located in different physical memory locations.

The Point new expression creates a point by sending the new message to the Point class. There are several ways to create a point:

- Point new creates a point (*nil, nil*), a point in which both x and y are left uninitialized and therefore have the nil value. All classes in Pharo understand the new message. Except when explicitly prohibited, an object is created by sending new to the class.

- `Point x: 5 y: 10` creates a point (5, 10). This expression sends the message `#x:y:` to the class `Point`, with 5 and 10 as the arguments. The `Point` class defines the class method named `#x:y:`. The difference between using `new` and `x:y:` to create a point is that the latter allows you to create and initialize a point with a given value for `x` and `y`.
- `2 @ 3` sends to the object `2` the message named `@` with the argument `3`. The effect is the same as with `Point x: 2 y: 3`, which is to create the point (2, 3).

A class may have its own way to create objects. For example, a point is not created the same way that a color is created. Creating an object is also commonly called "instantiating a class" and an object is often referenced as an "instance".

A class is an object factory and an object is necessarily created from a class. As discussed, objects interact each other by sending messages. An object can understand messages corresponding to methods defined in its class, and methods defined in the chain of superclasses.

Creating Classes

A class is both a factory and abstraction of objects. You need to create classes as soon as you want to bundle logic and data together. In Pharo, a program is necessarily structured as a set of interacting classes.

Pharo comes with a few thousand classes, and they need to be structured in some way so as not to overwhelm programmers. Pharo, like most programming languages, offers the notion of the *package*, which is essentially a container of classes. A class belongs to a package. A good practice is to create a package to contain the classes you will define:

1. Open a system browser from the top menu.
2. Right-click the top left list panel (see Figure 3-3) and define a package called `TweetsAnalysis`.

3. Create the Tweet class. A class is created by adding the following to a code browser:

```
Object subclass: #NameOfSubclass
    instanceVariableNames: ''
    classVariableNames: ''
    package: 'TweetsAnalysis'
```

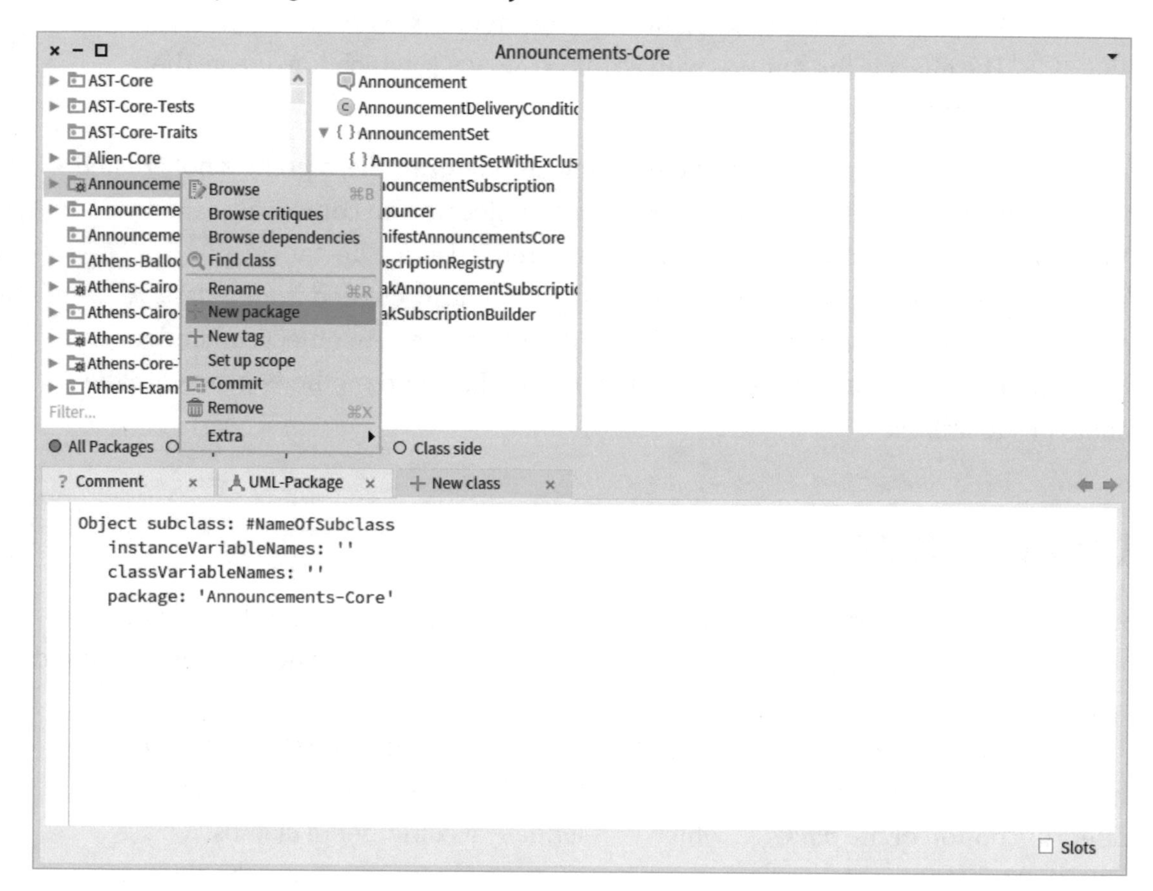

Figure 3-3. *Creating a new package*

The *system browser* is a standard tool in Pharo that allows you to browse and edit the source code. The previous code is a template for class creation. The `NameOfSubclass` text has to be replaced with the name of the class you want to create. After the `instanceVariableNames:` keyword, you need to provide the instance variables, and after `classVariableNames:`, provide the class variables. A class variable is visible from instance methods and class methods. Note that we do not use class variables in this book, so the `classVariableNames:` field will always be left empty.

In the Tweet example, you will model a tweet as an object having content, a sender, and a date. You could therefore define the following Tweet class:

```
Object subclass: #Tweet
    instanceVariableNames: 'content sender date'
    classVariableNames: ''
    package: 'TweetsAnalysis'
```

Write (or simply copy and paste this definition from the GitHub link given at the beginning of the chapter) this class definition in the lower of in the code browser, right-click the pane, and select Accept. In Pharo jargon, accepting a class definition means creating and compiling the class.

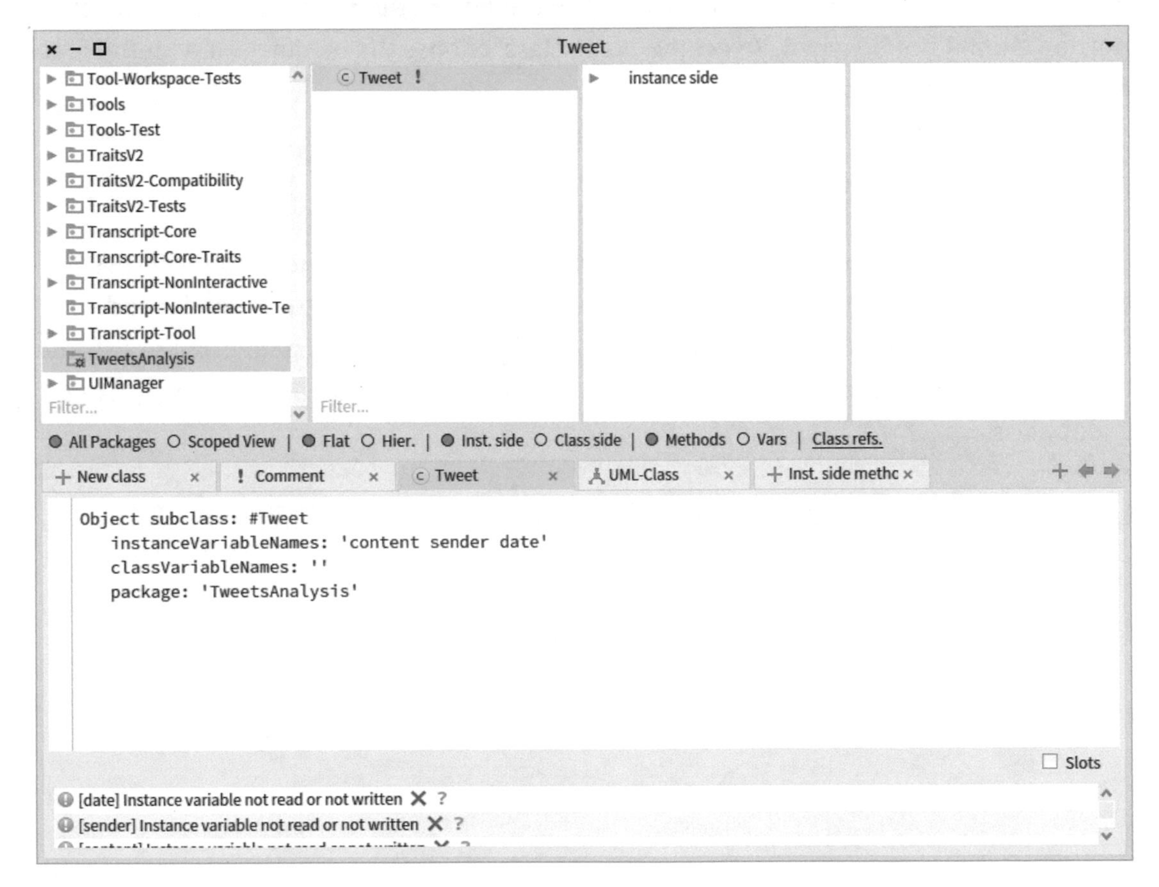

Figure 3-4. *Creating a new class*

Figure 3-4 shows the system browser after having created the Tweet class, which is contained in the TweetsAnalysis package.

Creating Methods

A *method* is an executable piece of code. A method is composed of instruction statements that carry out a computation. No methods have been defined so far. In this case, if you want to define meaningful tweet objects, you need to modify the values of the variables defined in the Tweet class. To access the date of a tweet, you define the method with the following source code:

```
Tweet>>date
    ^ date
```

The source code in this book uses the convention of preceding the method name with the class name. The code defines the date method in the Tweet class. The Tweet>> portion should not be typed. Select the Tweet class, choose the instance side Method tab, and enter the following:

```
date
    ^ date
```

You can now compile the code you just entered by pressing Cmd+S (on macOS) or Ctrl+S (on Linux and Windows platforms). Figure 3-5 illustrates the state of the code browser after compiling the method. The upper-right part of the system browser lists the methods defined in the selected class. As you can see, the method date is now part of the Tweet class.

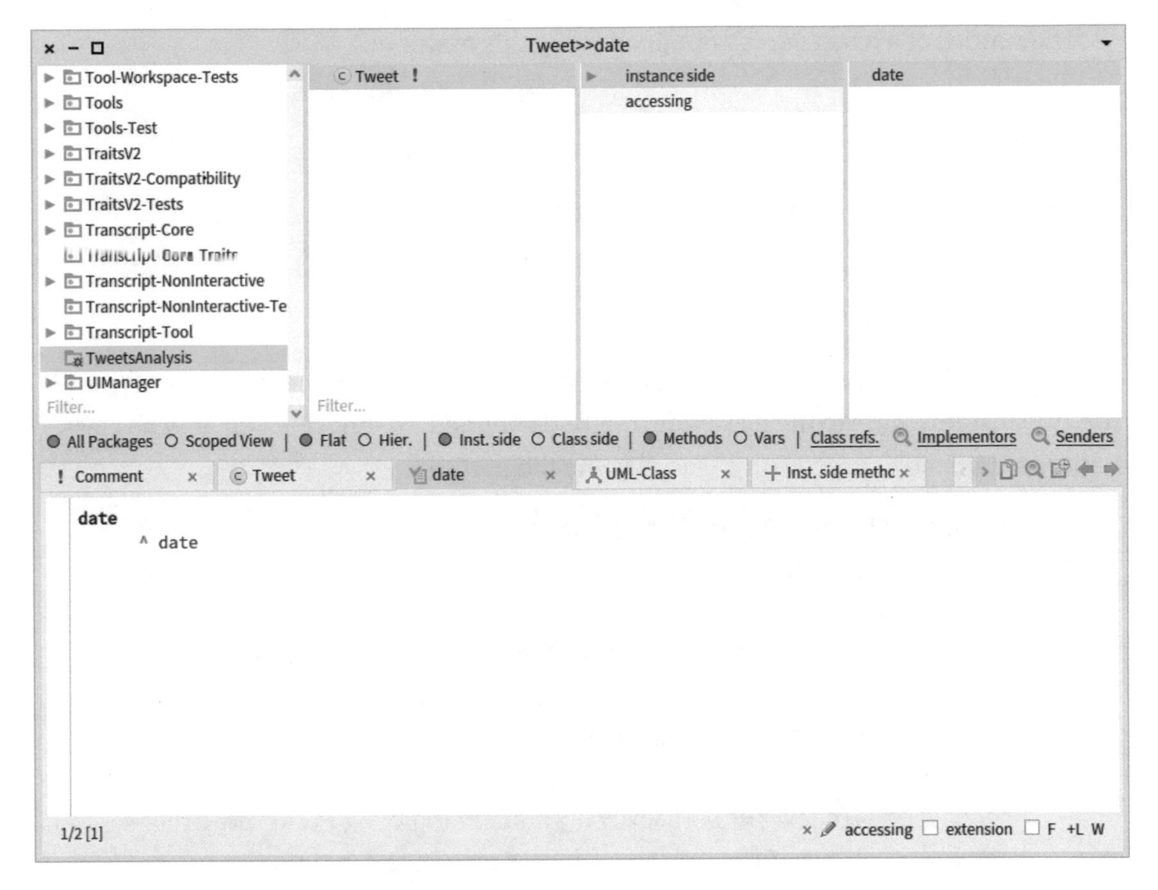

Figure 3-5. *Creating a new method*

The `date` method is useful for fetching the date from a tweet. The `date:` method allows you to set a date on a `Tweet` object:

```
Tweet>>date: aDate
    date := aDate
```

Similarly, the contents of a tweet can be retrieved using:

```
Tweet>>content
    ^ content
```

A tweet's content can be set using:

```
Tweet>>content: aContent
    content := aContent
```

The sender of a tweet can be obtained using this method:

```
Tweet>>sender
    ^ sender
```

A sender can be set in a Tweet object using:

```
Tweet>>sender: aSender
    sender := aSender
```

Click the Class Side button in the system browser. It switches the system browser from the instance side to the class side: methods defined on that side will now be class methods, invocable directly on a class. Define the method:

```
Tweet class>>createFromURL: urlAsString
    "Method to be defined on the CLASS side"
    | content lines sender date |
    content := (ZnEasy get: urlAsString) contents readStream.
    lines := content contents lines collect: [ :l |
        | firstCommaIndex secondCommaIndex |
        firstCommaIndex := l indexOf: $,.
        secondCommaIndex := l indexOf: $, startingAt: (firstCommaIndex + 1).
        sender := l copyFrom: 1 to: (firstCommaIndex - 1).
        date := l copyFrom: (firstCommaIndex + 1) to: (secondCommaIndex - 1).
        content := l copyFrom: (secondCommaIndex + 1) to: l size.
        { sender . date . content }
    ].
    ^ lines collect: [ :line |
        Tweet new
            sender: line first;
            date: line second;
            content: line third ]
```

The createFromURL: method fetches the CSV file we have prepared for that example. Note that the comments in the code are double quoted in Pharo. The file contains 1,000 random tweets. It does a simple parsing of the content by splitting it using commas. Next, you can define the method:

```
Tweet class>>createFromExample
    "Method to be defined on the CLASS side"
```

```
| url |
url := 'http://agilevisualization.com',
       '/AgileVisualization/tweets.csv'.
^ self createFromURL: url
```

The provided URL is an example to illustrate the example in this chapter. You can open the `http://agilevisualization.com/AgileVisualization/tweets.csv` URL in a web browser to see what the CSV file looks like. At this stage, evaluating the expression Tweet `createFromExample` returns a list of 1,000 tweet objects, each tweet describing an entry of the online CSV file. See Figure 3-6.

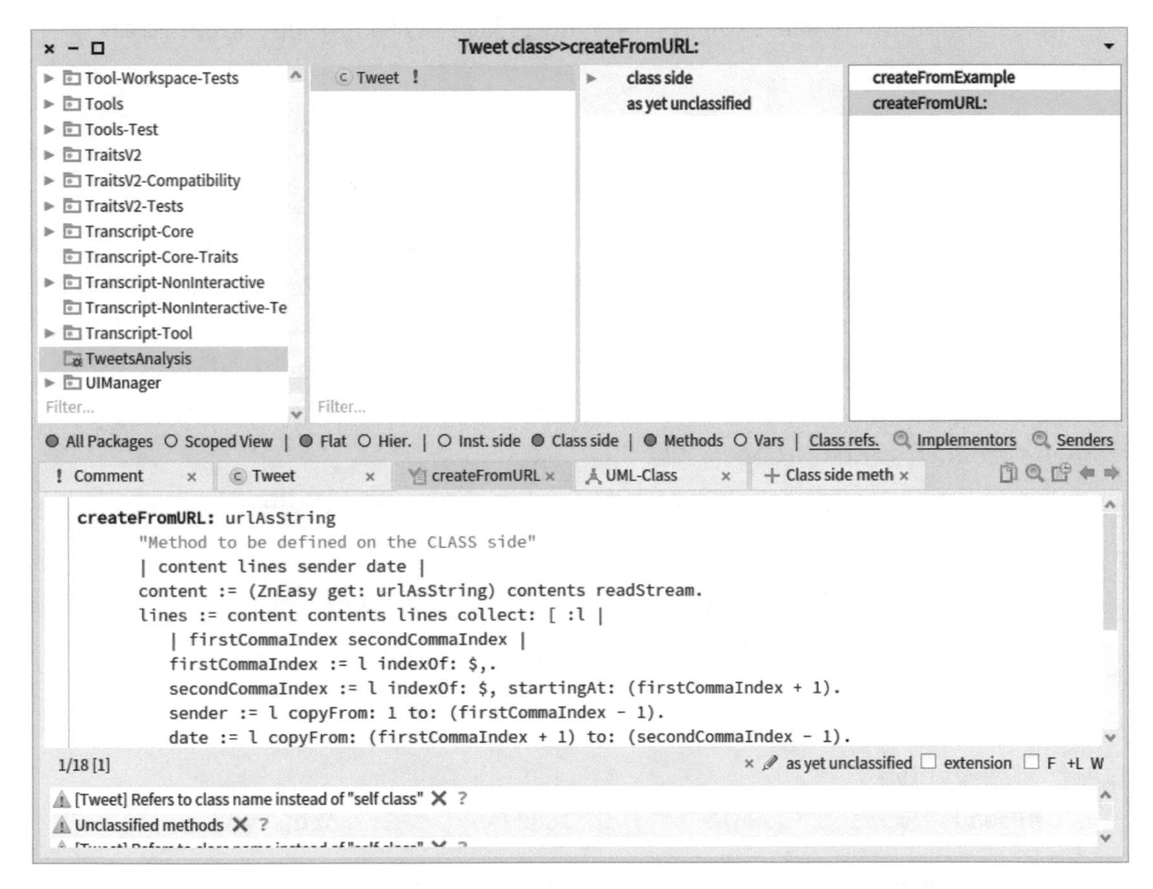

Figure 3-6. *The createFromURL: method, defined in the Tweet class*

You can define two new methods on the Tweet class. Switch to the instance side (i.e., select the Inst. side button in the system browser) and define the following two instance methods:

```
Tweet>>words
    "Return the list of words contained in a tweet"
    ^ self content substrings
```

The words method returns the list of words contained in a tweet. The words method uses substrings to return a list of words from a string, the content of the tweet. For example, the expression 'fox and dog'substrings returns #('fox''and''dog').

You can compare tweets using the method isSimilarTo:, which is defined as follows:

```
Tweet>>isSimilarTo: aTweet
    ^ (self words intersection: aTweet words) size >= 4
```

The isSimilarTo: method takes as an argument another tweet and returns true or false, depending on whether the tweet argument is similar to the tweet that receives the isSimilarTo: message. The notion of similarity used here is that two tweets are similar if they have at least four words in common. This is a simple and easy-to-implement heuristic, and it does not consider the actual semantics of the tweet; however, it is enough to build a structure between tweets.

Here's a simple way to determine whether a tweet conveys a positive feeling:

```
Tweet>>isPositive
    ^ #('great' 'cool' 'super' 'fantastic' 'good' 'yes' 'okay' 'ok')
    includesAny: self words
```

To determine if a tweet is negative, define this method:

```
Tweet>>isNegative
    ^ #('bad' 'worse' 'down' 'no') includesAny: self words
```

Again, this example uses very simple heuristics to identify the feeling conveyed in a tweet. The goal is to illustrate relevant and important aspects of Pharo, not to convey the state-of-the-art in sentiment analysis.

For this scenario, you need to have a meaningful textual representation of a tweet. Currently, if you print a tweet by typing Tweet new and pressing Cmd+P/Ctrl+P, you will obtain "a Tweet", which is not really useful in this situation. You will now define the method printOn: to produce an adequate textual representation:

```
Tweet>>printOn: str
    | whatToPrint |
    whatToPrint := self content
                    ifNil: [ 'Empty' ]
                    ifNotNil: [ self content ].
    str nextPutAll: whatToPrint
```

You now have some objects and a way to establish a relationship between them based on a simple heuristic for similarity. This is enough to visualize the tweets and look for some patterns. Open the Playground and type the following (see Figure 3-7).

```
tweets := Tweet createFromExample.
tweetShapes := RSCircle models: tweets forEach: [ :aCircle :aTweet | aTweet
isPositive ifTrue: [ aCircle color: Color green ].
 aTweet isNegative ifTrue: [ aCircle color: Color red ] ].
tweetShapes translucent.

c := RSCanvas new.
c addAll: tweetShapes.
RSLineBuilder line
    shapes: c nodes;
    color: Color gray translucent;
    withBorderAttachPoint;
    moveBehind;
    noBidirectional;
    connectToAll: [ :tweet |
            tweets select: [ :t | t isSimilarTo: tweet ] ].

c nodes @ RSPopup @ RSDraggable.
```

```
RSNormalizer size
    shapes: c nodes; from: 5; to: 15; normalize: [ :tweet | tweet content size ].

RSConditionalLayout new
 ifNotConnectedThen: RSGridLayout new;
 else: RSForceBasedLayout new;
 on: c nodes.
c @ RSCanvasController.
c open.
```

Figure 3-7. *Visualizing some tweets*

The visualization shows a large cluster of connected tweets. Tweets that are not similar to any others are on the left side, using a grid layout.

Block Closures

A *block closure* (also simply called "block") is a piece of code associated with an environment. A block is manipulable, as is any Pharo object, and can be provided as a message argument and be assigned to a variable. The expression `[ :value | value + 5 ]` is a block closure that takes one parameter and adds 5 to it. This block can be evaluated with an argument using the `value:` message. Consider the following code snippet:

```
b := [ :value | value + 5 ].
b value: 10. "Return 15"
b value: -5. "Return 0"
```

A block can accept zero, one, or more arguments. A block without any argument may be `[ 42 ]`, and is evaluated by simply sending `value` to it. For example, `[ 42 ] value` returns `42`. As you will soon see, this is particularly useful in control structures such as conditional statements. An example of a block with more than one argument could be:

```
b := [ :x :y | x + y ].
b value: 40 value: 2 "return 42"
```

Control Structures

Conditional statements are expressed using one of the following messages: `ifTrue:ifFalse:`, `ifTrue:`, or `ifFalse:`.

For example, the value `(42 > 5)ifTrue: [ 'Great value']` is evaluated to `'Great value'`. The argument provided to `ifTrue:` must have no argument.

The `(42 > 5)` expression evaluates to a boolean, to which the message `ifTrue:ifFalse:` is sent. This `ifTrue:ifFalse:` message takes two blocks as arguments. The first one is evaluated when the boolean receiver is `true`, and the second block is evaluated when the receiver is `false`.

Collections

A *collection* is a very common data structure and it is critical that you understand how to use them. As previously illustrated, the `#(23 42 51)` expression defines an array, which is an instance of the `Array` class. You can verify the class of an array by evaluating

#(23 42 51)class, which gives Array. This class, and its superclasses, have a large number of methods. Data transformation and data filtering are two common operations typically performed on a collection.

A transformation is realized using collect:. For example, #(23 42 51)collect: [:v | v > 30] returns #(false true true). The initial array of numbers is transformed as an array of booleans.

Filtering is carried out using select:. For example, #(23 42 51)select: [:v | v > 30] returns #(42 51). Both collect: and select: take a block as an argument. In the case of select:, the block has to evaluate to a boolean.

Pharo's collections are rich and expressive. You just saw the example of Array. Another useful collection is OrderedCollection, which represents an expandable collection. Elements may be added to and removed from an ordered collection during program execution. For example:

```
v := OrderedCollection new.
v add: 23.
v add: 42.
v add: 51.
shapes := v collect: [ :nb | RSBox new size: nb ].
RSVerticalLineLayout new alignCenter; on: shapes.
RSCanvas new
    addAll: shapes;
    yourself
```

This small script produces three squares lined up vertically. Another useful collection is Dictionary. A dictionary stores pairs of keys and values. For example, consider the following code snippet:

```
d := Dictionary new.
d at: #one put: 1.
d at: #two put: 2.
d at: #three put: 3.
```

The d at: #two expression returns the value 2 and the expression d at: #four raises an error.

Cascades

A *cascade* is expressed using the syntactical element and it allows you to send several messages to the same object receiver. For example, instead of writing:

```
v := OrderedCollection new.
v add: 23.
v add: 42.
v add: 51.
```

You can write:

```
v := OrderedCollection new.
v
    add: 23;
    add: 42;
    add: 51.
```

We this notation extensively when manipulating RSLineBuilder and RSNormalizer, as cascades considerably shorten the amount of code you need to provide.

A Bit of Metaprogramming

Pharo provides an expressive reflective API, which means you can programmatically get data about how Pharo code is structured, defined, and executed. Consider the RSShape methods size expression. This expression returns the number of methods that the RSShape class defines. The methods message is sent to the RSShape class, which is also an object in Pharo. This message returns a collection of the methods defined on the RSShape class. The size message is finally sent to that collection to obtain the number of methods defined in the RSShape class.

Many examples in Agile Visualization visualize software source code and use the reflective API. Visualizing source code is convenient because it does not need to fetch data from an external source of data, manipulating source code is trivial and very well supported in Pharo, and software source code is complex enough that it deserves to be visualized.

What Have You Learned in This Chapter?

This chapter gave a brief introduction to object-oriented programming. From now on, you should be able to understand Pharo syntax. I recommend a number of books to further discover the world of Pharo: *Pharo by Example* and *Deep Into Pharo,* both available for free from `https://books.pharo.org`.

Pharo is a beautiful, elegant, and simple language. It has a small and concise syntax, which makes it easy to learn. Its programming environment is also highly customizable.

Building a sophisticated visualization or any non-trivial software artifact often involves complex development. Mastering object-orientation is not strictly necessary in order to use Roassal and learn from through this book. However, having a good command of object-oriented programming will considerably alleviate your development and maintenance efforts.

Pharo offers a powerful meta-architecture. Do you remember that an object is created by sending the `new` message to a class? In Pharo, a class is also an object since you send `new` to it, as in the expression `Color new`. A class is therefore an object, itself an instance of another class, called a *metaclass*. And it does not stop here. A metaclass is also an object. Methods are also objects, each method being a collection of bytecodes. Many parts of Pharo are truly beautiful, but going into more detail is out of the scope of this book.

CHAPTER 4

Agile Visualization

Visualization is central to many domains, including data analysis and artificial intelligence. Nowadays, it has never been so easy to write a visualization. Most programming languages come with several libraries dedicated to visualizing data. As such, the technical aspects necessary to implement a visualization engine and to build a visualization are now largely understood. Despite all the experience gained by the software engineering community in building efficient visualization engines, there are still some challenges that have been poorly addressed globally. In particular, one of the challenges that needs to be properly addressed when designing a library to visualize data is being able to integrate visualizations in an existing production environment. Connecting and integrating one or more visualizations in a given environment is an important and non-trivial challenge. Agile visualization, as promoted by the Roassal visualization engine for the Pharo programming language, provides a solution to that problem, which is the topic of this book. This chapter develops the idea of Agile visualization.

All the code provided in this chapter is available at `https://github.com/bergel/AgileVisualizationAPressCode/blob/main/01-04-AgileVisualization.txt`.

Visualizing Classes as a Running Example

This chapter incrementally builds visualizations of a software component source code. From a data point of view, a software source code is a complex piece of data: source code does not easily fit in a `.csv` file and standard data analyses techniques cannot be run on software source code. Visualizing software source has many applications, ranging from software quality assessment to reengineering the software architecture. As such, visualizing software source code is a reasonable and non-trivial task.

Let's begin with a simple example. Consider the following code, executable within the Pharo Playground:

A. Bergel, *Agile Visualization with Pharo*, https://doi.org/10.1007/978-1-4842-7161-2_4

```
"The variable classes contains the classes we would like to visualize"
classes := Collection withAllSubclasses.

"A canvas is a container of graphical shapes"
c := RSCanvas new.

"Each class is represented as a box"
classes do: [ :aClass | c add: (RSBox new model: aClass) ].

"The width of each class indicates the number of variables defined in the class"
RSNormalizer width shapes: c shapes; from: 6; to: 20;
    normalize: #numberOfVariables.

"Height of each class represents the number of methods"
RSNormalizer height shapes: c shapes; normalize: #numberOfMethods.

"A class color goes from gray to red, indicating the number of lines of code"
RSNormalizer color shapes: c shapes;
    from: Color gray; to: Color red; normalize: #numberOfLinesOfCode.

"Vertical lines indicate the inheritance relationship"
RSLineBuilder orthoVertical
    canvas: c; withVerticalAttachPoint; color: Color lightGray;
    connectFrom: #superclass.

"Use a tree layout to adequately locate the classes"
RSTreeLayout on: c nodes.

"We make all the classes draggable and with a contextual popup window"
c nodes @ RSDraggable @ RSPopup.

"The whole visualization is zoomable, draggable, and shapes may be searched
in it"
c @ RSCanvasController.
```

After typing this code in the Playground, execute it and inspect its result using Cmd+G (macOS) or Ctrl+G (Windows/Linux). The result is shown in Figure 4-1.

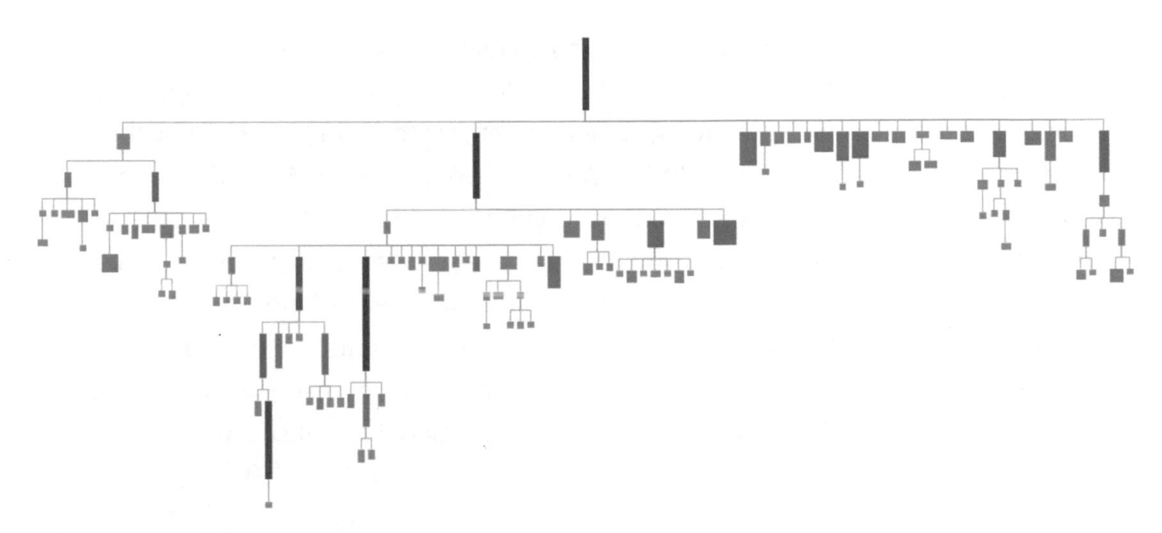

Figure 4-1. *Visualizing classes*

Figure 4-1 represents the class hierarchy of the `Collection` component of Pharo. The `Collection` component is an essential part of Pharo and consists of a set of classes to build collections, lists, and arrays of Pharo objects. The visualization uses a simple metaphor in which each box is a class. Lines indicate inheritance, and in particular, a superclass is located above its subclasses. The height of a box indicates the number of methods defined in the represented class, while the width represents the number of variables. The color of the boxes ranges from gray to red, indicating the number of lines of code of the represented class. A tall and large box indicates a class with many methods and is defined by many lines of code. Conversely, a small gray box indicates a class that doesn't have many methods or variables and has only a few lines of code. This visualization gives an overview of the source code distribution over the class hierarchy. It also enables you to spot exceptional entities (i.e., classes that are visually very different from other classes).

We will now detail the script. The `classes` variable refers to a set of Pharo classes. Each class is represented as an `RSBox` object, added to the `c` canvas. One important feature of Roassal is support of the connection between a graphical shape and the object model represented by the shape. This connection is expressed using the `model:` message, as in the expression `RSBox new model: aClass`. The use of the `model:` message makes the connection explicit between an arbitrary object (e.g., a class in this case) and the graphical shape.

To define the shape and color of each box, you use normalizers, which use the model object. A class in Pharo answers to many different messages. For example, the String numberOfMethods expression gives a number greater than 300, which is the number of methods defined in the String class. The number of methods defined in the String class depends on the considered version of Pharo. Similarly, you can send the numberOfVariables, numberOfLinesOfCode, superclass message to a class to obtain the number of instance variables, the number of lines of code, and the superclass, respectively. Lines are built between classes to indicate inheritance using a dedicated object, the RSLineBuilder. The classes, positioned in the canvas using the tree layout, are draggable and the class name appears as a popup when the mouse cursor hovers a class.

Example in the Pharo Environment

The provided code example looks relatively standard, and the same visualization can easily be built using any modern visualization engine. If you are familiar with a library such as D3.js or Matplotlib, you might say: "Well, I can do the same". This is true up to a point. The real benefit of Roassal is not the API. Beneath the surface, Roassal offers a number of features to make the visualization navigable and facilitate the integration within the Pharo environment.

A graphical shape in Roassal is linked to an object model. The simple expression RSBox new model: aClass makes a box a façade of a class, and this way of representing data is very different compared to the way other engines operate. By having the connection between a graphical shape and the object model explicit, you can click a shape to inspect the object model. In this case, you can click that box to open an Inspector on the clicked class. The Inspector provides a number of built-in visual representations for the inspected object.

You will now define a new visual representation of the call graph of methods defined in a class. Consider the following method defined on the class named Class, itself contained in the 'Kernel-Classes' package:

```
Class>>visualizeCallGraph
    "Visualize the call graph of the class' methods"
    | c eb shapes |
    shapes := RSBox models: self methods forEach: [ :box :cm |
        box color: #blue ].
```

```
    RSNormalizer size shapes: shapes; normalize: #numberOfLinesOfCode.
    shapes @ (RSPopup text: #selector) @ RSDraggable.
    c := RSCanvas new.
    c addAll: shapes.
    eb := RSLineBuilder arrowedLineWithOffset: 0.2.
    eb moveBehind.
    eb shapes: c nodes.
    eb canvas: c.
    eb withVerticalAttachPoint.
    eb connectToAll: #dependentMethods.
    RSTreeLayout on: c nodes.
    shapes @ RSHighlightable withLines.
    ^ c @ RSCanvasController
```

This method can be invoked on any class. For example, inspecting the `String visualizeCallGraph` expression visualizes the call graphs of `String`'s methods. The `visualizeCallGraph` method first creates a box for each method. Lines are then added to indicate invocations between methods. You now have to notify the Inspector framework to call `visualizeCallGraph` when inspecting a class. The Inspector framework can easily be extended with the following method:

```
Class>>>inspectorVisualization
    <inspectorPresentationOrder: 90 title: 'Callgraph'>
    ^ SpRoassal3InspectorPresenter new
        canvas: self visualizeCallGraph;
        yourself
```

The `inspectorVisualization` method creates a visualization showing the call graph when a class is inspected.

```
Class>>>inspectorVisualizationContext: aContext
    "Remove the evaluator pane"
    aContext withoutEvaluator
```

Now that a visualization is defined for classes by extending `Class`, clicking classes rendered by the very first script of the chapter shows details of the call graph. This connects the two visualizations. Similarly, a method can be selected to display a wide range of information, including the byte code and source code. Consider the example in Figure 4-2.

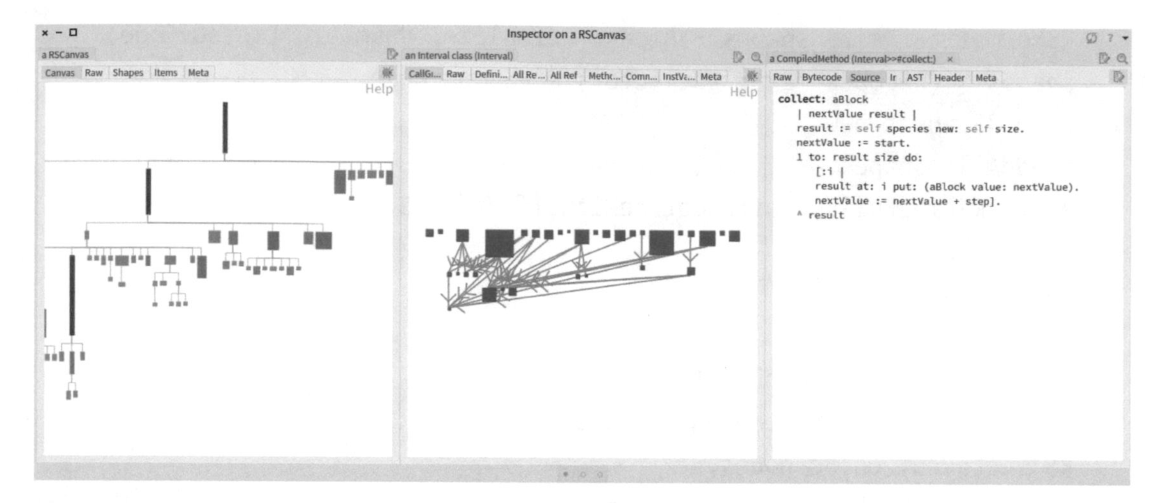

Figure 4-2. *Visualizing classes*

In Figure 4-2, the most-left pane represents the visualization of the classes given in the initial script. In this first visualization, the `Interval` class is selected. The selection displays the method call-graph at the center. In the call-graph, the `collect:` method is selected, and it shows the method source code on the right side.

To summarize, you have independently built two visualizations. The Inspector framework exposes them in a very convenient way while supporting their navigations. You can jump from the first one to the second one, even their scripts are independent.

Closing Words

Imagine being empowered with a tool to easily prototype visualizations and to embed them in your work environment. If the cost of building visualizations and their integration is low enough, you would likely write more visualizations more frequently. As soon as visualizations are easy to write and use in daily tasks, visualization will pop up like mushrooms in your production environment. We define the notion of *agility* in data visualization as lowering the production cost of a visualization, and lowering the cost of integrating visualization in a production environment. Agility turns a beautiful picture in a tool to enable practitioners to act upon a valuable domain and enjoy immediate feedback.

Roassal3 is the result of more than ten years of hard work. The Pharo and Smalltalk communities have played an important role in shaping the API and identifying the key aspects of what Agile visualization is.

What Have You Learned in This Chapter?

This chapter illustrated connecting two independent visualizations, a cornerstone of Agile visualization. In particular, the chapter covered:

- Visualizations can be defined for any object, by simply defining visualization methods on a class. A visualization is hooked into the Inspector framework using the pragma `inspectorPresentationOrder:title:` and using the `SpRoassal3InspectorPresenter` class. Several visualizations can be defined in a class.
- Clicking a shape opens a new pane in the Inspector and offers a visualization defined in the class of the selected object.

CHAPTER 5

Overview of Roassal

This chapter begins the second part of the book, which focuses on the core of Roassal and includes an overview of its main components. This chapter goes one step further than the previous chapters by listing the essential components of Roassal.

All the code provided in this chapter is available at `https://github.com/bergel/AgileVisualizationAPressCode/blob/main/02-01-Roassal.txt`.

Architecture of Roassal

The Roassal framework is composed of a number of macro components:

- *Shapes*: Visual graphical elements (e.g., box, line, polygon, bitmap, SVG element).
- *Canvas*: A container of shapes. A canvas is typically represented as a window.
- *Events*: In order for a graphical shape to respond to a user event (e.g., mouse movement or a keyboard keystroke), events need to be adequately managed.
- *Interactions*: A pluggable, composable, and customizable behavior that is usually triggered from a user interaction.
- *Normalizers*: Shapes may be shaped and colored in a meaningful way to represent attributes and properties of the objects represented by the shapes.
- *Layouts*: Shapes may be located to accommodate the overall physical distribution.

A. Bergel, *Agile Visualization with Pharo*, https://doi.org/10.1007/978-1-4842-7161-2_5

- *Inspector integration:* Pharo offers a sophisticated Inspector to enable interactive and visual inspections and navigation.
- *Animations*: Built using dedicated abstractions.

The chapter briefly describes and gives an illustration of each of these components. The next chapters explains each of these macro components separately.

Shapes

A *shape* is a visual element meant to be displayed. Roassal offers numerous shapes, which are highly configurable. The `RSShape` class is the root of all the Roassal shapes, and a shape is typically created by instantiating a subclass of `RSShape`. For example, a box is defined by instantiating `RSBox`.

A colored box with a thick border may be defined as follows:

```
box := RSBox new
            size: 80;
            color: #yellow;
            border: (RSBorder new color: #blue; width: 10).
```

Colors are set using the `color:` message and the argument is either a color object (e.g., `Color blue` or `Color r: 0 g: 0 b: 1.0`) or a symbol (e.g., `#yellow`). When a symbol is provided, it is converted to a color object using `Color>>colorFrom:`, as in the expression `Color colorFrom: # yellow`.

Although most of the configuration may be directly carried out by sending an object with a literal (e.g., `size: 80`), part of the shape's configuration may involve dedicated objects, such as `RSBorder`.

Shapes may be subject to geometrical transformations, including translation, rotation, and scaling. Shapes may also answer some particular events triggered by a user.

A shape may represent a Pharo object, called a *model object.* Models are useful when applying interactions and applying normalization, as described in this chapter. A model is set using the `model:` message, for example: `RSBox new model: Dictionary new`. In this case, the dictionary provided as a model is visually represented by the box shape. Shapes and colors of the box may reflect some properties of the dictionary. If a canvas is represented in an Inspector, clicking the box will inspect the dictionary.

Canvas

A canvas is essentially a container of shapes. Shapes may be added to a canvas using #add:. For example (see Figure 5-1):

```
box := RSBox new
            size: 00;
            color: #yellow;
            border: (RSBorder new color: #blue; width: 10).
canvas := RSCanvas new.
canvas add: box.
canvas open
```

Figure 5-1. *A yellow box with a blue border*

Sending the open message to a canvas creates a window in the Pharo environment that renders the canvas. A canvas can be open different ways. For example, the `openWithTitle:` method is a variant of open that lets you provide a title. Execute the following to see the result (see Figure 5-2):

```
RSCanvas new
    add: (RSBox new size: 100; color: #blue);
    openWithTitle: 'This is a blue box'
```

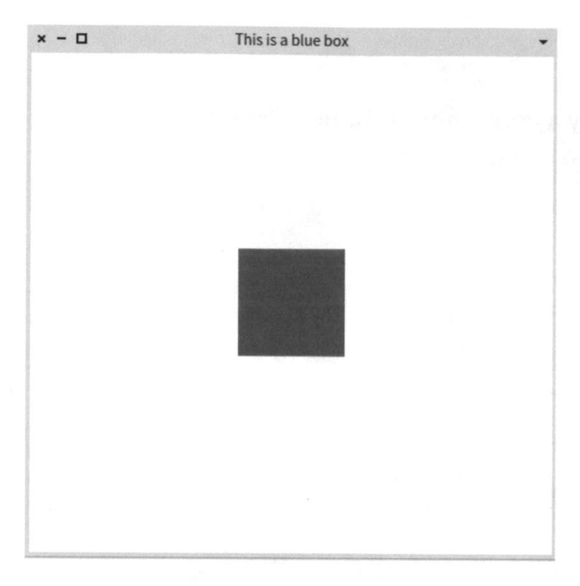

Figure 5-2. *Providing a title to a window*

The background of a canvas can be set using `#color:` This is handy when you want to have a canvas that matches the user interface color theme. Consider the following:

```
box := RSBox new
            size: 80;
            color: #yellow;
            border: (RSBorder new color: #blue; width: 10).
canvas := RSCanvas new.
canvas color: (UITheme current backgroundColor).
canvas add: box.
canvas open
```

This script opens a canvas whose color depends on the installed Pharo color theme.

The window size may be too small to render the content of the canvas. Scroll bars and zooming facilities can be set on a canvas using the `RSCanvasController` interaction. Consider the following example (see Figure 5-3):

```
canvas := RSCanvas new.

1 to: 1000 do: [ :i |
    canvas add: (RSBox new size: 10).
].
```

```
RSGridLayout on: canvas shapes.

canvas @ RSCanvasController.
canvas open
```

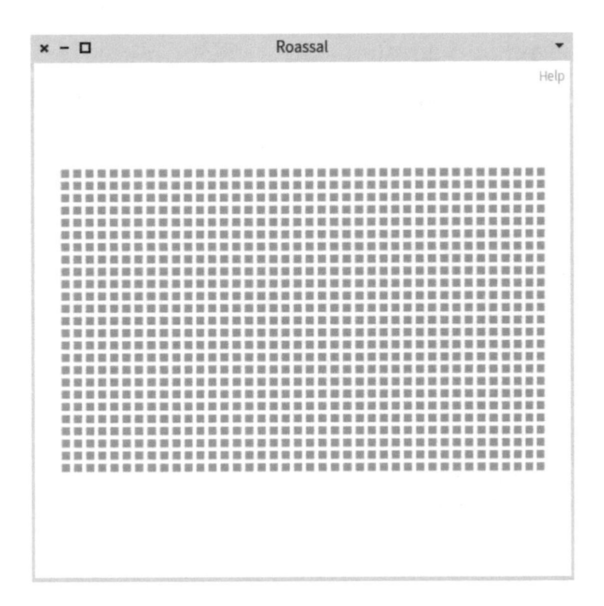

Figure 5-3. *Many boxes*

You can now zoom in and out by pressing the I and O keys on your keyboard. When the shapes are outside the window, scroll bars appear to ease the navigation. Scrolling may be achieved by pressing the arrows. Speed scrolling is obtained when you press the Shift key along with an arrow key.

Events

Being able to interact with shapes is crucial to making a visualization dynamic and reactive to user interaction. A human user can interact with shapes by emitting events through mouse clicks, mouse movements, and key strokes. A shape can have a particular behavior, which is then triggered when a specific event occurs.

Being able to interact with a single and a group of shapes is crucial when building a data visualization. Events may be useful to highlight, transform, and filter some shapes to support an effective visualization.

The RSEvent class is the root all the events supported by Roassal. An event is modeled as an instance of a subclass of RSEvent.

Consider the following script:

```
canvas := RSCanvas new.
labelAdd := RSLabel text: 'Add shapes'.
labelAdd @ RSHighlightable red.
canvas add: labelAdd.

labelAdd setAsFixed.
labelAdd translateTopLeftTo: 0 @ 0.

r := Random new.
canvas when: RSMouseClick do: [ :evt |
    canvas shapes do: #remove.
    canvas signalUpdate ].

labelAdd when: RSMouseClick do: [ :evt |
    100 timesRepeat: [
        shape := RSBox new
                        size: (r nextInt: 50);
                        color: Color random translucent;
                        translateTo: (r nextInt: 500) @ (r nextInt: 500).
        canvas add: shape ].
    "refresh the window"
    canvas signalUpdate.
    "make sure everything fits on screen"
    canvas zoomToFit ].

canvas open
```

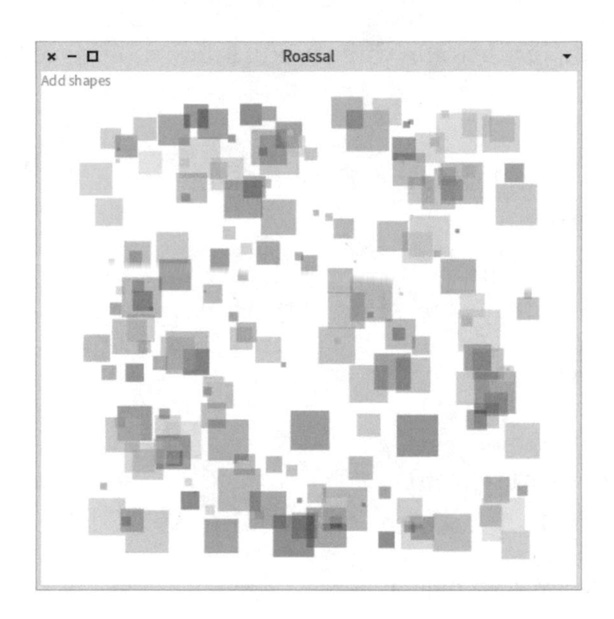

Figure 5-4. *Callbacks on shapes*

The previous script adds a label on the top-left corner of the canvas. Clicking the label has the effect of adding many boxes of random size, color, and position. The canvas is cleared when you click the background. See Figure 5-4.

This example illustrates a callback added to `labelAdd`. Clicking with the mouse on a shape emits an `RSMouseClick` event. The callback (i.e., the code to be executed in case of an event) is set using the `when:do:` message. The first argument is the class of the event and the second argument is a one-argument block that expresses the behavior to perform when the event occurs. The block accepts an event as an argument, from which the related shape and canvas can be accessed.

A callback is set on the canvas itself. Clicking the canvas clears it by removing all its shapes.

As you have seen, most of the events describe actions made by users (e.g., `RSMouseClick` and `RSMouseDragEnd`), some other events describe state changes of a shape (e.g., `RSShapeAddedEvent` and `RSHighlightEvent`). Such state-related events are useful when defining high-level interactions, as you will learn shortly.

Interaction

The `Roassal3-Interaction` package proposes several expressive interactions that are both customizable and composable.

The following example illustrates three useful interactions (see Figure 5-5):

```
canvas := RSCanvas new.

someColors := { Color red . Color blue . Color green }.
shapes := RSGroup new.
someColors do: [ :aColor |
    shapes add: (RSEllipse new size: 30; model: aColor; color: aColor) ].

canvas addAll: shapes.

shapes @ RSDraggable.
shapes @ RSLabeled highlightable.
shapes @ (RSPopup text: [ :aColor | 'My color is ', aColor asString]).

RSHorizontalLineLayout on: shapes.

canvas @ RSCanvasController.
canvas open.
```

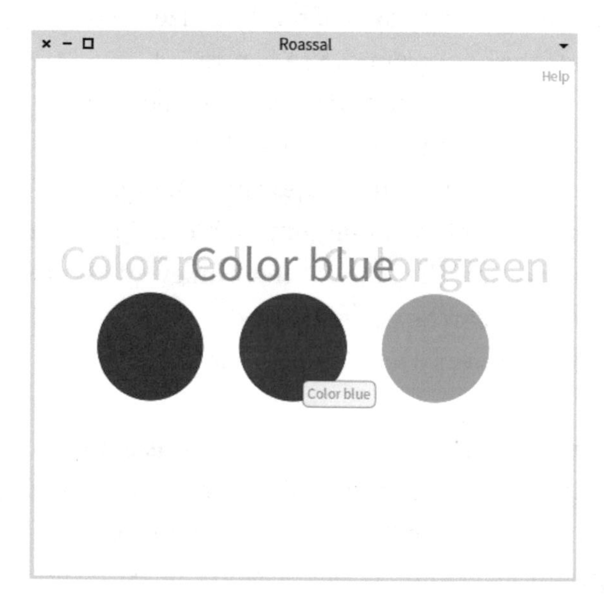

***Figure 5-5.** Composing interactions*

The `someColors` variable is a collection of three color objects. For each of these colors, a circle is created using the `RSEllipse` shape. Each shape has a color and a model object matching the color it represents. It is important to realize that in this example, we have some objects (kept in the variable `someColors`) that are visually represented by some shapes (kept in `shapes`). The `shapes` variable is a Roassal group in which shapes are added.

After being added to a canvas, three interactions are set on these shapes. The `RSDraggable` integration makes the shapes draggable using the mouse. The `RSLabeled` interaction adds a label on top of the shape that describes the model object. The label is highlighted when the mouse hovers over the colored circle. A popup is a contextual window that appears when the mouse hovers over a colored circle. The text shown in the popup can be modified using a block that expects the model represented by the shape.

Roassal offers a wide range of interactions—`RSDraggable`, `RSLabeled`, and `RSPopup` are the most common.

Normalizer

It is useful to map numerical values to visual attributes, including size and color. Comparing visual elements often involves normalizing some particular values and mapping them to properties of the shapes. Consider the following example (see Figure 5-6):

```
values := #(20 30 40 50 10).

c := RSCanvas new.
shapes := RSEllipse models: values.
shapes @ RSPopup.
c addAll: shapes.

RSNormalizer size
    shapes: shapes;
    from: 20;
    to: 30;
    normalize.

RSNormalizer color
    shapes: shapes;
    normalize.
```

```
RSGridLayout new gapSize: 40; on: shapes.
shapes @ RSLabeled.

c @ RSCanvasController.
c open
```

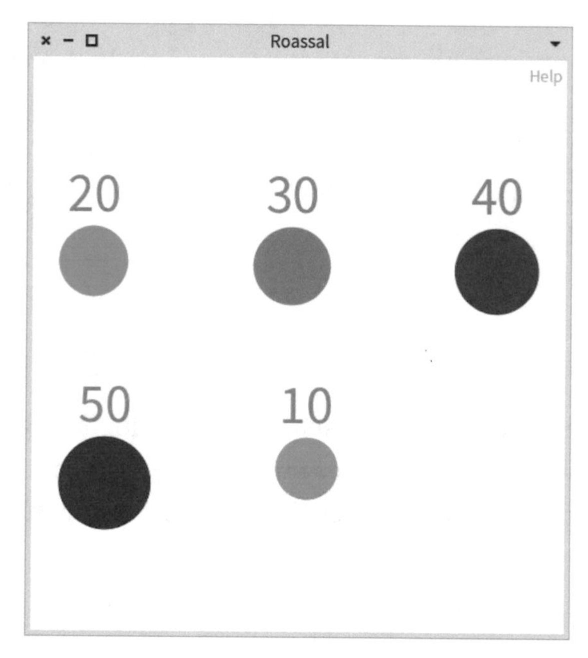

Figure 5-6. *Normalizing values*

The script begins by defining a set of numerical values kept in the `values` variable. Subsequently, a collection of shapes using `RSEllipse class>>models:` is added. A popup is set to each member of the collection. The size and color of the shapes are normalized using the `RSNormalizer` class.

The essential class for normalization is `RSNormalizer` and it offers numerous ways to normalize shapes.

Layouts

Roassal offers numerous layouts to exploit the space resource offered by a canvas. Adequately locating shapes in a two-dimensional plane may highlight some properties or a hierarchy among the represented shapes.

Consider the following example (see Figure 5-7):

```
values := #(20 10 5 30 24 32).
links := { 20 -> #(10) . 5 -> #(24 . 20) . 10 -> #(32 30) } asDictionary.
canvas := RSCanvas new.
shapes := RSEllipse models: values.
canvas addAll: shapes.
shapes @ RSPopup.
RSNormalizer size
    shapes: shapes; normalize.

RSLineBuilder line
    shapes: shapes;
    connectToAll: [ :nb | links at: nb ifAbsent: [ #() ] ].

RSTreeLayout on: shapes.
canvas @ RSCanvasController.
canvas open
```

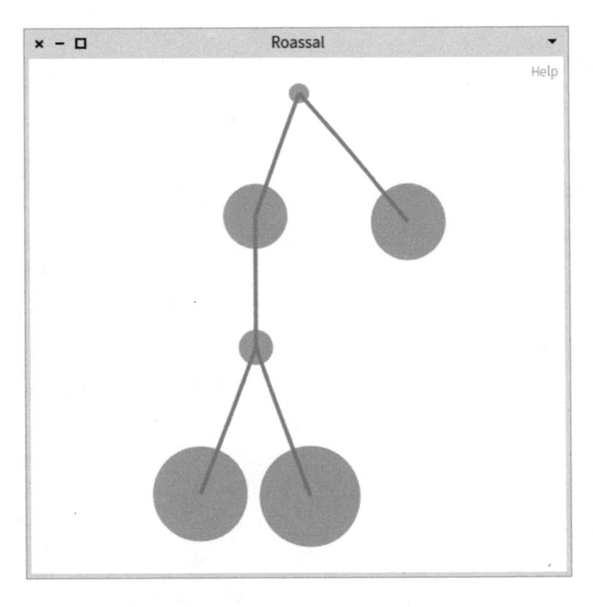

Figure 5-7. *Layout example*

The script above uses a RSLineBuilder to define lines between shapes. The RSTreeLayout class arranges the shapes as a tree. Roassal offers numerous layouts to arrange shapes. Replacing RSTreeLayout with RSClusterLayout, RSRadialTreeLayout, or RSFlowLayout will produce a different arrangement.

Inspector Integration

Pharo offers a flexible framework to support specific object inspectors. Roassal provides the necessary tools to build object-specific visualization. You simply need to define one or two methods on a class to make a visualization available to its instances.

In the Playground, type the expression 1 / 3 and click the green Do It button (or use Cmd+G/Alt+G). A new pane appears that shows the default object inspector. Evaluating the 1 / 3 expression produces a fraction object, an instance of the class Fraction, which is defined in the Kernel-Numbers package. You can easily define a visual representation of a fraction by defining the following method on the Fraction class:

```
Fraction>>inspectorView
    <inspectorPresentationOrder: 90 title: 'Graphic'>
    | c numeratorShape denominatorShape line |
    c := RSCanvas new.
    numeratorShape := RSLabel model: numerator.
    denominatorShape := RSLabel model: denominator.
    c add: numeratorShape.

    line := RSLine new
                from: 0 @ 0;
                to: ((numeratorShape width max: denominatorShape width) +
                20) @ 0.
    c add: line.

    c add: denominatorShape.

    RSVerticalLineLayout new gapSize: 2; alignCenter; on: c shapes.
    c @ RSCanvasController.

    ^ SpRoassal3InspectorPresenter new
        canvas: c;
        yourself
```

This method produces a visual representation of a fraction. The visualization appears when a fraction is inspected in the Inspector. Evaluate the expression `1 / 3` once more in the Playground. You should obtain the result shown in Figure 5-8.

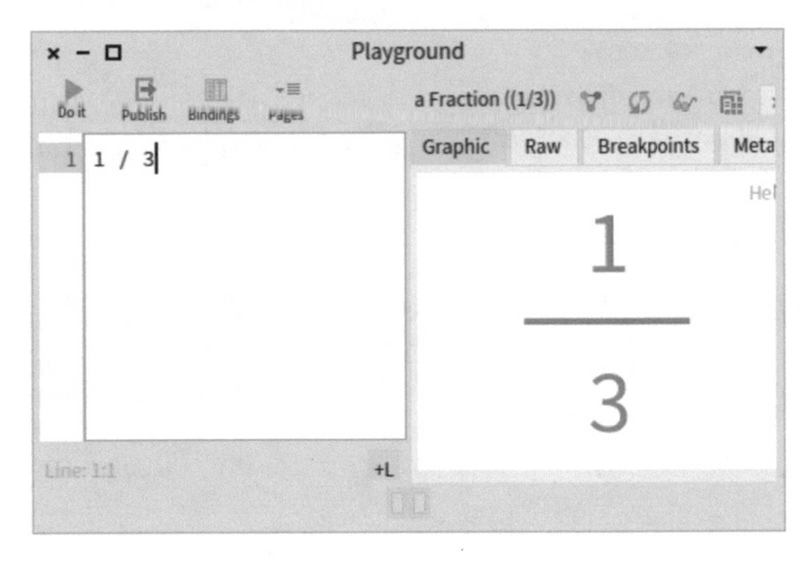

Figure 5-8. *Visualizing a fraction*

Animation

Roassal offers an expressive infrastructure to build and execute animations. The `Roassal3-Animation` package contains all the classes related to animations. A canvas can build an animation by simply sending `newAnimation` to it. Various methods are useful to configure the numerous parameters that define an animation. Consider the following script (see Figure 5-9):

```
c := RSCanvas new.
r := Random seed: 42.

100 timesRepeat: [
    p := ((r nextInt: 500) - 250) @ ((r nextInt: 500) - 250).
    shape := RSCircle new color: Color random translucent.
    shape translateTo: p.

    c add: shape.
```

```
    c newAnimation
        easing: RSEasing bounce;
        from: p;
        duration: ((r nextInt: 5) / 2) second;
        to: p + (0 @ 100);
        on: shape set: #position:;
        when: RSAnimationEndEvent do: [ :evt | shape remove ].
].
c
```

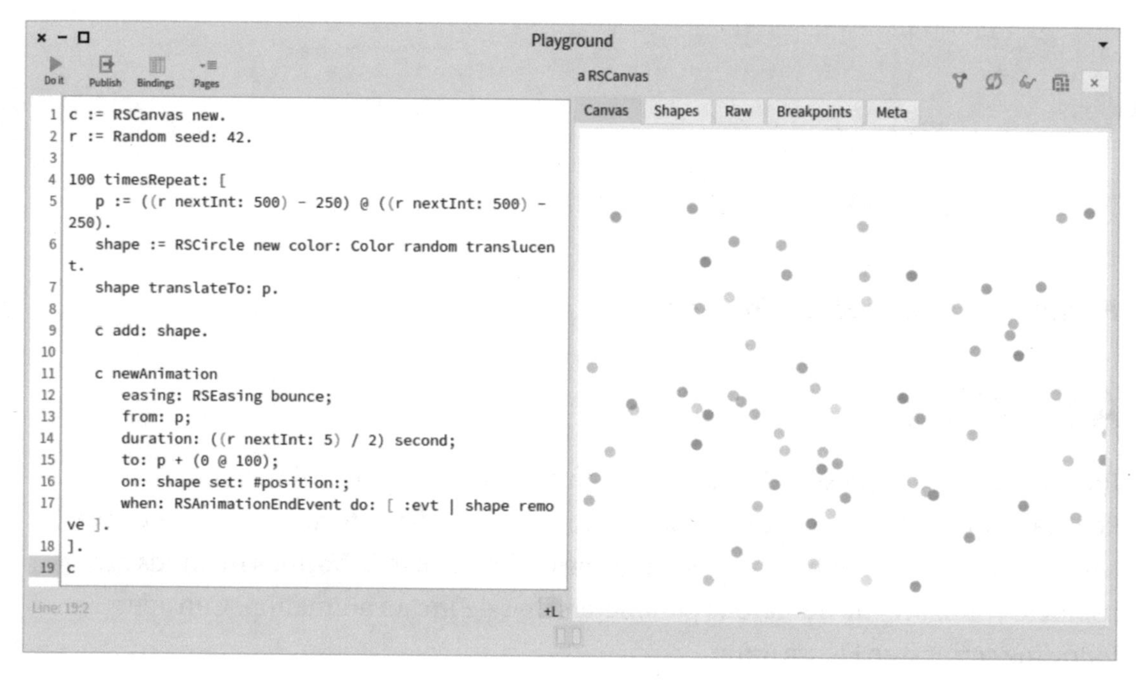

Figure 5-9. *Example of an animation*

What Have You Learned in This Chapter?

This chapter provided a broad overview of Roassal and shallowly described the essential components. In particular, it provided a short and concise example of each macro component in Roassal. The subsequent chapters will review in detail the essential points introduced in this chapter.

CHAPTER 6

The Roassal Canvas

A visualization in Roassal is presented in a canvas. The canvas acts as a placeholder for graphical shapes, and shapes may be added to and removed from a canvas. The canvas is modeled using the RSCanvas class, and most of the operations related to a canvas are expressed using messages.

All the code provided in this chapter is available at https://github.com/bergel/AgileVisualizationAPressCode/blob/main/02-02-Canvas.txt.

Opening, Resizing, and Closing a Canvas

Creating and opening a canvas is central when using Roassal. A canvas is created using the RSCanvas new expression. A canvas is open by sending open to it. For example, the RSCanvas new open expression opens an empty, white window.

Shapes may be added to a canvas using add:. The following code simply creates a canvas and adds a box shape to it:

```
c := RSCanvas new.
c add: RSBox new.
c open
```

Simply sending open to it opens a new window showing the canvas content, as illustrated here.

It frequently happens that a visualization is built in an incremental fashion, by executing many similar but different codes. Each small increment in a visualization involves opening and closing windows. This could be cumbersome as it requires you to manually close windows (either using the mouse or using keyboard shortcuts such as Cmd+W/Ctrl+W) is rather cumbersome when there are multiple increments. For example, when you try to modify the shape's color and size in the script. Each execution opens a new window that you need to close in order to not keep your screen clear of

A. Bergel, *Agile Visualization with Pharo*, https://doi.org/10.1007/978-1-4842-7161-2_6

windows. Replacing open with openOnce makes the window reusable. Each execution of the script displays the result in the same window. Under the hood, a new window is created each time a canvas opens, but this trick is completely transparent.

Opening a canvas returns an object describing the Pharo window. Many things can be done with these windows, including resizing and closing. Consider the following code snippet:

```
c := RSCanvas new.
c add: (RSBox new color: Color red; size: 200).
win := c open.
win extent: 800 @ 300.
```

The open method returns a SystemWindow object. Sending extent: 800 @ 300 resizes the window. The default size of the window is 500 @ 500.

A convenient way to maximize the visualization in a window is to use the zoomToFit message. Consider the following example:

```
c := RSCanvas new.
c addAll: (RSLabel models: (1 to: 9)).
RSGridLayout on: c shapes.
c zoomToFit.
c open
```

Sending the zoomToFit message fits the canvas content to the window. This method is useful for maximizing the space provided by the Pharo host window.

Simply sending the delete message removes the window from the Pharo environment. For example, the delete message removes the window that contains the canvas. Consider the following example:

```
c := RSCanvas new.
lbl := RSLabel text: 'Click on me to close the window'.
lbl fontSize: 30.
lbl @ RSHighlightable red.
c add: lbl.
win := c open.
win extent: 600 @ 100.
lbl when: RSMouseClick do: [ :evt | win delete ]
```

Executing this code snippet opens an horizontal window with a label in it. Clicking the label removes the window.

Camera and Shapes

By default, shapes are subject to camera movement, which means that when you move the camera attributes (e.g., its position), the shapes appear to be affected. This behavior can be overridden by sending the setAsFixed message to the shape. Consider the following example:

```
c := RSCanvas new.

fixedLbl := RSLabel new text: 'Fixed label'; fontSize: 20.
c add: fixedLbl.
fixedLbl setAsFixed.
fixedLbl translateTopLeftTo: 10 @ 10.

lbl1 := RSLabel new text: 'Movable label'; fontSize: 20.
c add: lbl1.

lbl2 := RSLabel new text: 'Try moving the background'; fontSize: 20.
c add: lbl2.
lbl2 translateBy: 0 @ -30.

c @ RSCanvasController.
c open
```

The example defines three labels—fixedLbl, lbl1, and lbl2. The first label is fixed, which means that dragging the canvas does not affect it. The other two labels move when the camera is moved. The c @ RSCanvasController instruction makes the canvas' camera draggable.

Fixed shapes can be used as a simple way to create a menu toolbar. Consider the following example (see Figure 6-1):

```
c := RSCanvas new.

addLabel := RSLabel text: 'Add shapes'.
clearLabel := RSLabel text: 'Clear'.
changeColorLabel := RSLabel text: 'Change color'.
c add: addLabel; add: clearLabel; add: changeColorLabel.

addLabel setAsFixed.
clearLabel setAsFixed.
changeColorLabel setAsFixed.
```

```
labels := RSGroup withAll: { addLabel . clearLabel . changeColorLabel }.
labels @ RSHighlightable red.
RSHorizontalLineLayout on: labels.

addLabel when: RSMouseClick do: [ :evt |
    10 timesRepeat: [
        shape := RSEllipse new size: 30.
        r := Random new.
        shape translateTo: (r nextInt: 500) @ (r nextInt: 500).
        c add: shape.
        c zoomToFit.
        c signalUpdate ] ].

clearLabel when: RSMouseClick do: [ :evt |
    c shapes do: #remove.
    c signalUpdate ].

changeColorLabel when: RSMouseClick do: [ :evt |
    c shapes do: [ :s | s color: Color random translucent ].
    c signalUpdate ].

c open
```

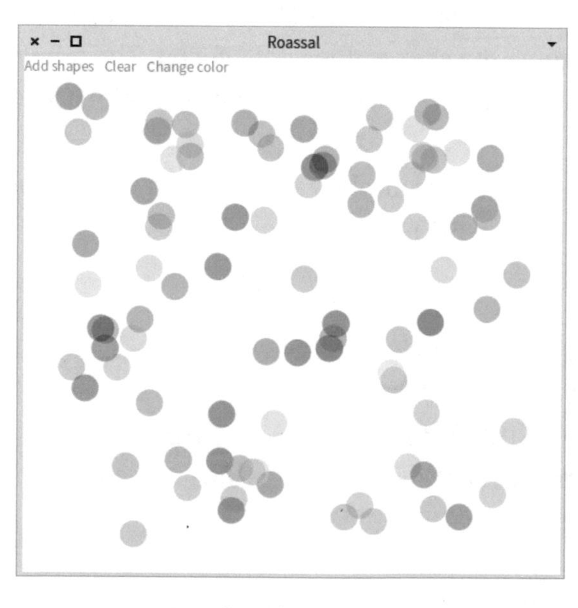

***Figure 6-1.** Defining a menu using fixed shapes*

Three fixed labels are defined. They can be highlighted using the `labels @ RSHighlightable defaultRed` expression. A callback for the `RSMouseClick` event is defined on each label. When the canvas content is modified, the `zoomToFit` and `signalUpdate` messages are sent to force a resize of the canvas and a refresh of the window by the Pharo window manager. Note that the three fixed labels are not affected by `zoomToFit`, thus mimicking the feel of an actionable menu.

Virtual Space

A canvas has an infinite two-dimensional virtual space. Each shape contained in a canvas has a position expressed as a point. To illustrate how the virtual space is divided, consider the following example (see Figure 6-2):

```
points := { { 'Center' . 0 @ 0 } .
                { 'Top' . 0 @ -150 } .
                { 'Top left' . -150 @ -150 } .
                { 'Top right' . 150 @ -150 } .
                { 'Bottom' . 0 @ 150 } .
                { 'Bottom left' . -150 @ 150 } .
                { 'Bottom right' . 150 @ 150 } }.
canvas := RSCanvas new.
points do: [ :tupple |
    point := RSCircle new.
    point translateTo: tupple second.
    point @ (RSLabeled text: tupple first, tupple second asString).
    canvas add: point.
].

canvas add: (RSLine new dashed; color: Color red; from: 0 @ -200; to: 0  @ 200).
canvas add: (RSLine new dashed; color: Color red; from: -200 @ 0; to:
200  @ 0).

canvas @ RSCanvasController.
canvas open
```

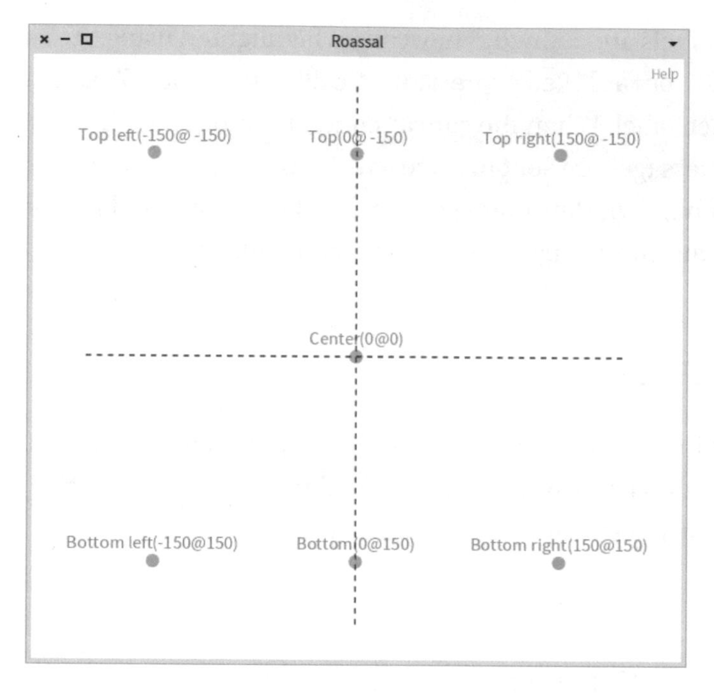

Figure 6-2. *Coordinate system used by the canvas*

Positions of shapes are expressed in *scaled pixels,* meaning that if no zooming is applied by the camera, distances are expressed in pixels. A camera object can be obtained from a canvas by sending `camera` to the canvas. The scaling factor can be manually set using the expression `canvas camera scale: 1`. The `scale:` method expects a float number. Providing a number smaller than 1 gives the effect of zooming out, while a number greater than 1 gives the effect of zooming in.

The position of a shape is defined as the position of the shape center. Consider a shape defined as `RSCircle new`. This circle has a diameter of 10 pixels. You can verify this by evaluating `RSCircle new extent`, which returns `10.0 @ 10.0`. The shape is framed within a rectangle called the *encompassing rectangle.* The top-left corner of the encompassing rectangle is (−5, −5), verifiable by evaluating `RSCircle new encompassingRectangle topLeft`.

Shape Order

Shapes are rendered onscreen in the same order they are added, and adding a shape may hide previously added shapes. The order may be explicitly modified by simply sending pushBack to a shape, which as the effect of moving back when being rendered. Consider the example (see Figure 6-3):

```
c := RSCanvas new.
redShape := RSCircle new size: 50; color: #red.
blueShape := RSCircle new size: 50; color: #blue.
blueShape translateBy: -25@0.
c add: redShape; add: blueShape.
blueShape pushBack.
c zoomToFit.
c open
```

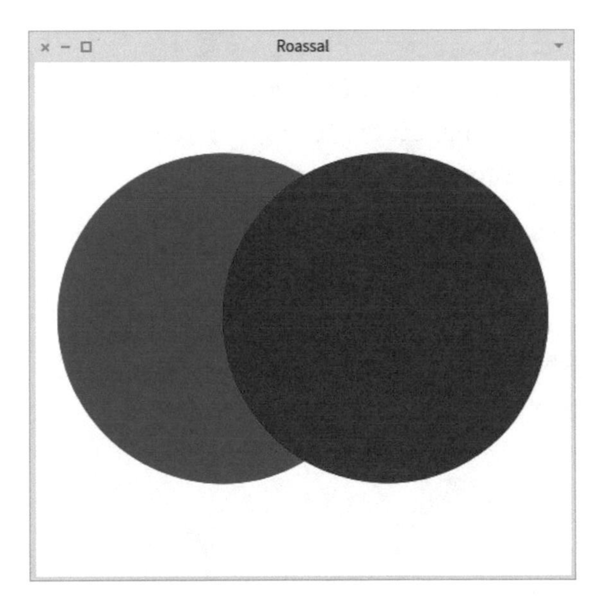

Figure 6-3. *Pushing back some shapes*

In this code, redShape is added before blueShape, which has the effect of making the red circle drawn before the blue circle. The position and the size of the circles make the blue shape overlap the red one. Since the blue circle is added before the red circle, the red circle hides part of the blue one.

However, the order of the shape rendering may be modified by pushing a shape to the back or to the front. The `blueShape pushBack` expression drawn the blue circle first. The `pushFront` method can also be employed.

Canvas Controller

By default, a canvas does not respond to any action triggered by the end-user. It is very convenient to make the canvas react to particular actions. It frequently happens that a user wants to move the canvas, search for particular shapes, and perform some zooming action. The canvas controller fulfills this need. Consider the following example (see Figure 6-4):

```
c := RSCanvas new.

r := Random new.
100 timesRepeat: [
    size := r nextInt: 50.
    shape := RSEllipse new size: size; model: size.
    shape translateTo: (r nextInt: 500) @ (r nextInt: 500).
    shape color: Color random translucent.
    c add: shape
].

c shapes @ RSPopup.
RSFlowLayout on: c shapes.
c @ RSCanvasController.
c open
```

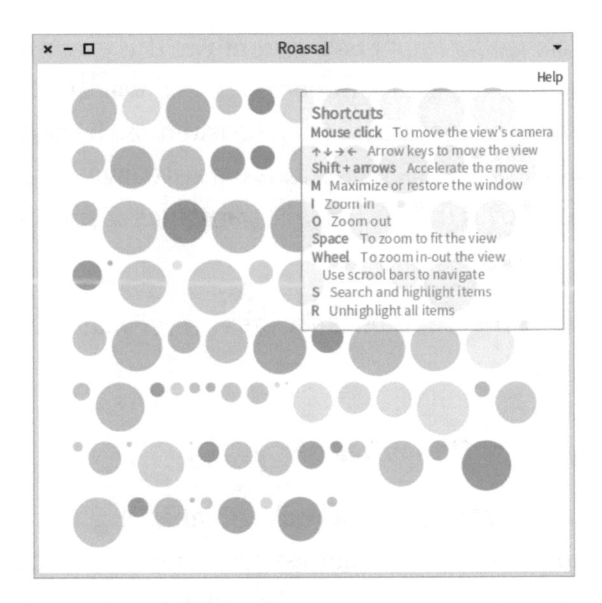

Figure 6-4. *Help offered by the canvas controller*

The `c @ RSCanvasController` expression adds a controller to the canvas. The canvas controller offers a set of options that may be activated through the keyboard or the mouse, in particular:

- The canvas may be dragged using the mouse.
- The canvas may be translated using the arrow keys; pressing Shift speeds up the translation.
- Scrollbars are also available to precisely control the camera position.
- The mouse wheel controls the zooming level.
- When open, the entire visualization is visible in the displayed window.
- A search facility is offered to search for particular elements matching a pattern. Pressing the key S opens a window asking for a textual pattern. Shapes that have a model whose textual representation matches the pattern are highlighted.
- Shapes that were previously matched in a search can get their original color by pressing R.

Setting the canvas controller happens frequently in this example, as you might already noticed. The options currently supported in a canvas are listed when activating the Help menu item. The canvas controller can be extended by adding a new option, simply by creating a subclass of the `RSAbstractControlCanvas` class. You will build a simple interaction that moves the camera to a particular shape. First, define the following class:

```
RSAbstractControlCanvasInteraction subclass: #MyFocusOnCanvas
    instanceVariableNames: ''
    classVariableNames: ''
    package: 'MyControllerExtension'
```

This example assumes that you have a package called `MyControllerExtension`. It's best to avoid defining classes whose name begin with RS, as that is reserved for the core of Roassal, and not its extension. Define the following method on the class you just created:

```
MyFocusOnCanvas>>renderLegendOn: lb
    lb text: 'F' description: 'Focus on an element by moving the camera'.
```

Overriding the `renderLegendOn:` method is useful to create the Help menu item. The method takes a `RSLegend` object as an argument. Sending the `text:descriptions:` message and add a label and its accompanying description. Feel free to browse the `RSLegend` class to see an exhaustive list of the methods that may be involved in the `lb` object.

Another important method for the controller extension is `onShape:`, which defines the behavior to be executed when the interaction is activated. Define the method:

```
MyFocusOnCanvas>>onShape: aCanvas
    aCanvas
        when: RSKeyUp do: [ :evt |
            evt keyName = 'F'
                ifTrue: [ self askAndFocusOn: evt canvas ] ]
```

The method essentially invokes the askAndFocusOn: method when the user presses the key F. The main method to define is as follows:

```
MyFocusOnCanvas>>askAndFocusOn: aCanvas
    "Ask for a shape to search"
    | modelToSearchAsString shapeToFocusOn |
    modelToSearchAsString := UIManager default
                                        request: 'Enter an object textual
                                        description'
                                        initialAnswer: 'E.g., 42'.

    "Exit if nothing was entered"
    (modelToSearchAsString isNil or: [ modelToSearchAsString isEmpty ])
    ifTrue: [ ^ self ].

    shapeToFocusOn := aCanvas shapes
                        detect: [ :s | s model asString = modelToSearchAs
                        String ]
                        ifNone: [ ^ self ].
    shapeToFocusOn @ RSBlink.
    aCanvas camera translateTo: shapeToFocusOn position.
    aCanvas signalUpdate.
```

First, the method asks for a textual description of the shape to be searched. The method then searches for the shape, makes it blink, and moves the camera on it.

You are now all set. You can try the extension of the canvas controller with the following example:

```
c := RSCanvas new.

r := Random new.
100 timesRepeat: [
    size := r nextInt: 50.
    shape := RSEllipse new size: size; model: size.
    shape translateTo: (r nextInt: 500) @ (r nextInt: 500).
    shape color: Color random translucent.
    c add: shape
].
```

```
c shapes @ RSPopup.
RSFlowLayout on: c shapes.
c @ (RSCanvasController new addInteractionClass: MyFocusOnCanvas).
c open
```

You can search for a shape by pressing the key F and entering the textual description of the object to search for. This simple example illustrates how actions can be added to a canvas controller.

Converting a Canvas to a Shape

A canvas is a closed and self-contained entity. Being able to compose canvases is therefore an appealing way to build a more complex visualization. As a simple example to illustrate the principle of composing canvases, consider the following code (see Figure 6-5):

```
canvas := RSCanvas new.
5 timesRepeat: [
    tmpCanvas := RSCanvas new.
    someShapes := RSCircle models: (1 to: 50).
    tmpCanvas addAll: someShapes.
    RSGridLayout on: tmpCanvas shapes.
    tmpCanvas shapes color: Color random translucent.
    canvas add: tmpCanvas asShape.
].

RSGridLayout on: canvas shapes.
canvas @ RSCanvasController.
canvas open
```

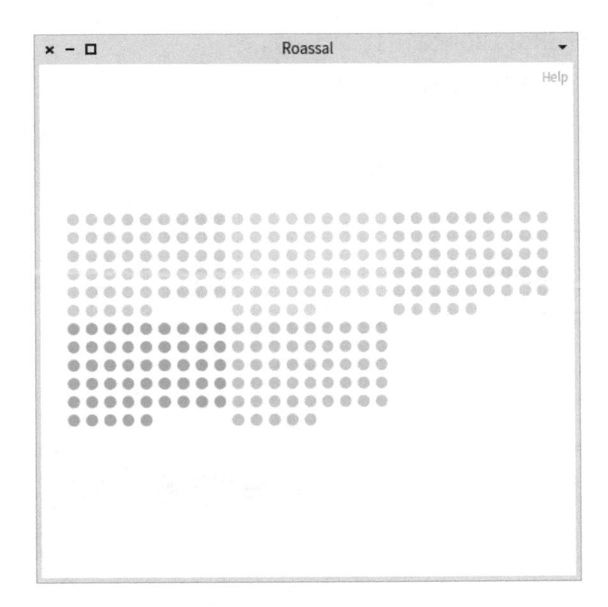

Figure 6-5. *Composing canvases*

Sending the asShape message to a canvas returns a composed shape, which can be added in a new canvas. A more sophisticated example is shown in the following code (see Figure 6-6):

```
g1 := RSChart new.
p1 := RSLinePlot new.
p1 y: #(5 10 3 -4 -5 15).
g1 addPlot: p1.
g1 addDecoration: (RSXLabelDecoration new title: 'Time'; offset: 0 @ 10).
g1 addDecoration: (RSYLabelDecoration new title: 'Value'; offset: -20 @ 0).
g1 addDecoration: (RSHorizontalTick new).
g1 addDecoration: (RSVerticalTick new).
g1 build.

g2 := RSChart new.
p2 := RSLinePlot new.
p2 y: #(2 10 50 -20 40 30 25 12 4).
g2 addPlot: p2.
g2 addDecoration: (RSXLabelDecoration new title: 'Time'; offset: 0 @ 10).
g2 addDecoration: (RSYLabelDecoration new title: 'Value'; offset: -20 @ 0).
```

```
g2 addDecoration: (RSHorizontalTick new).
g2 addDecoration: (RSVerticalTick new).
g2 build.

c := RSCanvas new.
c add: (g1 canvas asShape).
c add: (g2 canvas asShape).
RSHorizontalLineLayout on: c shapes.
c @ RSCanvasController.
c open
```

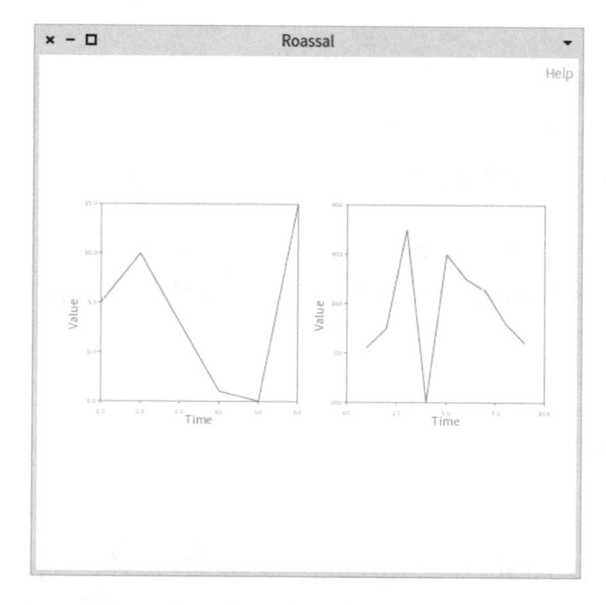

Figure 6-6. *Composing graphs by converting canvases into composed shapes*

The `RSChart` class builds a canvas with a chart in it. The canvas can be accessed by sending `canvas` to a chart. The example builds two graphs kept in the `g1` and `g2` variables. The canvas of each chart is converted to a shape and added to the `c` canvas.

A canvas also answers to the `asShapeModel:` message to attach a model to the composite shape returned by `asShape`. Attaching a model is useful for adding lines, for example. Consider this new iteration of the running example (see Figure 6-7):

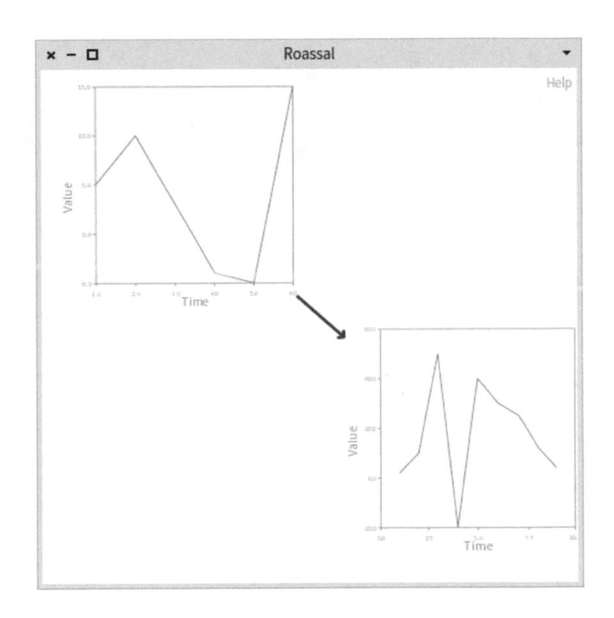

***Figure 6-7.** Adding an arrowed line between composed canvases*

```
g1 := RSChart new.
p1 := RSLinePlot new.
p1 y: #(5 10 3 -4 -5 15).
g1 addPlot: p1.
g1 addDecoration: (RSXLabelDecoration new title: 'Time'; offset: 0 @ 10).
g1 addDecoration: (RSYLabelDecoration new title: 'Value'; offset: -20 @ 0).
g1 addDecoration: (RSHorizontalTick new).
g1 addDecoration: (RSVerticalTick new).
g1 build.

g2 := RSChart new.
p2 := RSLinePlot new.
p2 y: #(2 10 50 -20 40 30 25 12 4).
g2 addPlot: p2.
g2 addDecoration: (RSXLabelDecoration new title: 'Time'; offset: 0 @ 10).
g2 addDecoration: (RSYLabelDecoration new title: 'Value'; offset: -20 @ 0).
g2 addDecoration: (RSHorizontalTick new).
g2 addDecoration: (RSVerticalTick new).
g2 build.
```

```
c := RSCanvas new.
c add: (g1 canvas asShapeModel: 1).
c add: (g2 canvas asShapeModel: 2).
RSHorizontalLineLayout on: c shapes.

"We make the graphs draggable"
c shapes @ RSDraggable.

lb := RSLineBuilder arrowedLine.
lb withBorderAttachPoint.
lb canvas: c.
lb useAssociation: 1 -> 2.

c @ RSCanvasController.
c open
```

The canvas obtained from `g1` is a composite shape representing the number 1, and the canvas from `g2` is associated with the value 2. These two numbers are then useful when building lines using a `RSLineBuilder`.

Converting a canvas to a shape comes with a few limitations. In particular, interactions that may be directly associated with a canvas are removed when converting. For example, the controller of the canvas is forgotten when the canvas is converted to a shape.

Turning a canvas into a shape is useful when you want to separately compose defined visualization that were not initially made to be composed. As you will later learn when reading about the `RSCompositeShape` class, Roassal offers complementary ways to compose and structure visualizations.

Events

Similar to a shape, a canvas can trigger events and have callbacks. Consider the following example:

```
canvas := RSCanvas new.

lbl := RSLabel new.
canvas add: lbl.
```

```
canvas when: RSEvent do: [ :evt |
    lbl text: evt asString.
    canvas signalUpdate.
].

canvas open
```

Actions triggered by a user—including moving the mouse, resizing the window, pressing a key, and using the mouse wheel—are associated with a particular event. In the previous example, the canvas has one callback that is executed at each event emitted by the canvas.

What Have You Learned in This Chapter?

This chapter covered the canvas, which is an essential element of Roassal. In particular, you learned that:

- A canvas is a container of shapes.
- Shapes may be fixed or not, to be subject to the canvas translation.
- Canvases can be composed.
- A controller offers relevant options to ease the navigation within a visualization.

CHAPTER 7

Shapes

Shapes are visual elements meant to be added to a canvas. A shape may be configured using various aspects, including color, border, line thickness, and many other parameters. This chapter details a number of shapes supported by Roassal.

All the code provided in this chapter is available at https://github.com/bergel/AgileVisualizationAPressCode/blob/main/02-03-Shapes.txt.

Box

A box is modeled with the RSBox class. For example, the following script adds a box to a canvas:

```
c := RSCanvas new.
box := RSBox new.
c add: box.
c open
```

Without any settings, as in this example, a box is gray and has a default size of 10 pixels per side. Its position is (0, 0), which corresponds to the center of the canvas when open.

The width and the height of a box can be set using `width:` and `height:`. Both expect a positive number (a float or integer). The `size:` message takes a number as a parameter and sets the height and the width. The `extent:` message, expecting a point, can be used to set the height and width in one message.

The color is set using `color:` and taking a `Color` object as an argument.

A. Bergel, *Agile Visualization with Pharo*, https://doi.org/10.1007/978-1-4842-7161-2_7

A box may have a corner radius, which gives it rounded corners. The `cornerRadius:` method expects a positive number greater than or equal to 0. Consider the following example, which sets a color, a corner radius, and a size (see Figure 7-1):

```
c := RSCanvas new.
r := Random seed: 42.
40 timesRepeat: [
    box := RSBox new width: (r nextInteger: 80); height: (r nextInteger: 80);
    cornerRadius: 10.
    box color: Color random translucent.
    box translateTo: (r nextInteger: 200) @ (r nextInteger: 200).
    c add: box.
].
c @ RSCanvasController.
c open
```

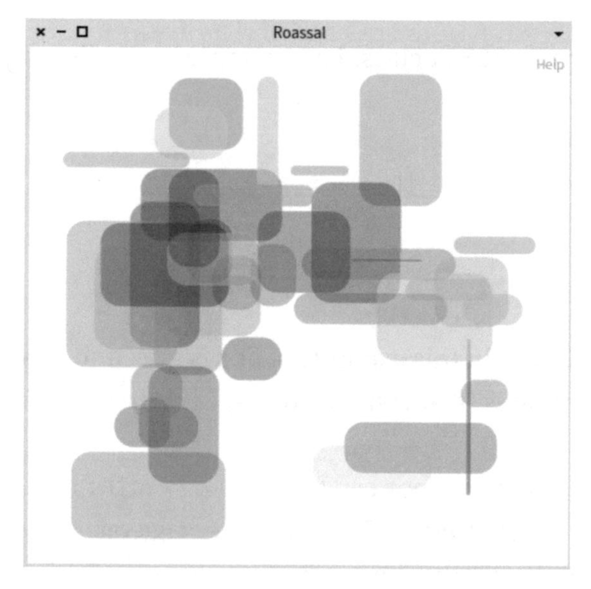

***Figure 7-1.** Some random boxes with a corner radius*

Circle and Ellipse

A circle is modeled using `RSCircle` and an ellipse is modeled using `RSEllipse`. Consider the following example (see Figure 7-2):

```
c := RSCanvas new.
(30 to: 150 by: 10) do: [ :nb |
    b := RSCircle size: nb.
    c add: b ].
RSFlowLayout on: c nodes.
c @ RSCanvasController.
c open
```

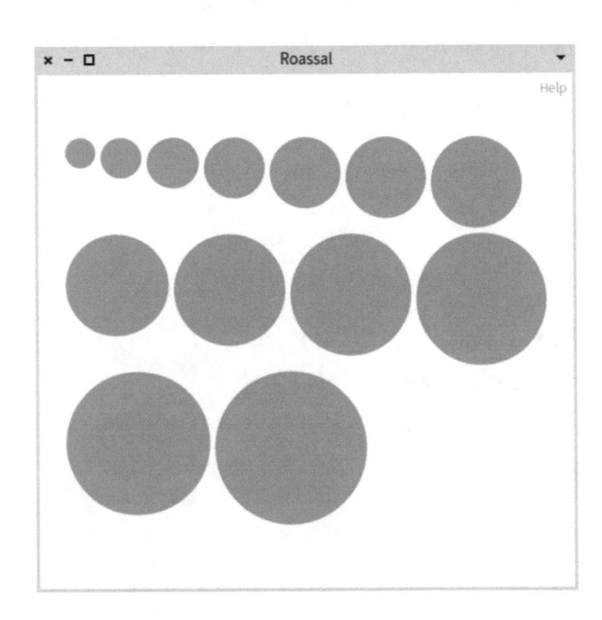

Figure 7-2. *Some circles*

The `RSCircle` and `RSEllipse` classes are similar to `RSBox`, since classes are subclasses of `RSBoundingShape`. Obviously, the `height:` and `width:` methods cannot be invoked on a circle. Furthermore, `RSCircle` and `RSEllipse` do not provide the method `cornerRadius:` as `RSBox` does.

Label

Labels are used to link visual elements to what they actually represent. Labels make a visualization interpretable and connected to a particular domain. The following example adds a label to a canvas:

```
c := RSCanvas new.
c add: (RSLabel text: 'Hello World').
c @ RSCanvasController.
c open
```

Labels are complex visual elements. For example, a label can have a particular font, a font size, and some visual attributes. The size of the font may be set using `fontSize:`. The following example uses a random font size for each word (see Figure 7-3):

```
words := String loremIpsum substrings.

c := RSCanvas new.
r := Random seed: 42.
words do: [ :w |
    label := RSLabel text: w.
    label fontSize: (r nextInteger: 30).
    label @ RSHighlightable red.
    c add: label ].

RSFlowLayout on: c shapes.
c @ RSCanvasController.
c open
```

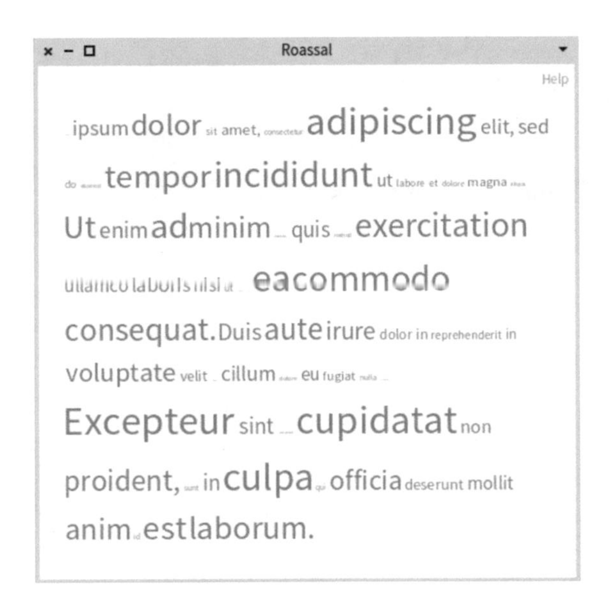

Figure 7-3. *Labels of different sizes*

The `loremIpsum` method defined on the class size of `String` gives a lorem ipsum, a classical placeholder text. Sending `substrings` to this text provides a list of words.

Most of the time, a shape cannot be considered a simple rectangular bounded area. In particular, a label has a baseline, which is useful when aligning textual shapes (see Figure 7-4). The baseline is the line that most letters "sit" on.

Figure 7-4. *Baseline of a label*

Some layouts of Roassal take the baseline into consideration. The flow and horizontal layout can be configured using `alignLabel` to align labels along their baseline, for example:

```
words := String loremIpsum substrings.

c := RSCanvas new.
r := Random seed: 42.
words do: [ :w |
    label := RSLabel text: w.
```

```
    label fontSize: (r nextInteger: 30).
    label @ RSHighlightable red.
    c add: label ].

RSFlowLayout new alignLabel; on: c shapes.
c @ RSCanvasController.
c open
```

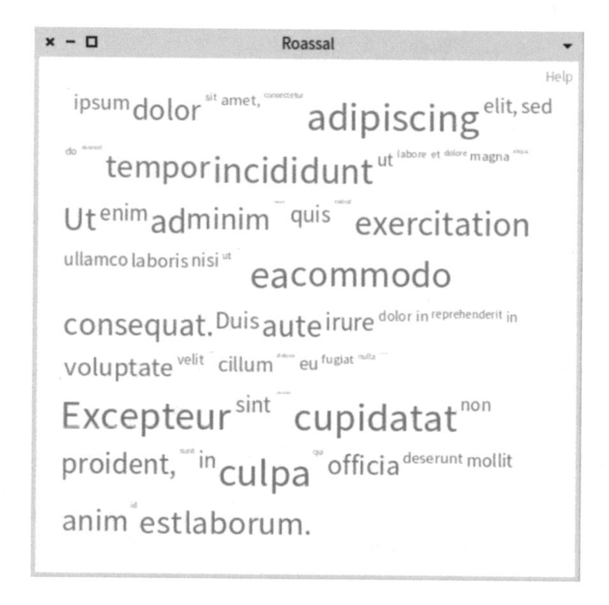

Figure 7-5. *Layout along the label baselines*

Roassal offers various strategies to compute the shape of a label. Such a strategy may have an impact when applying a layout or when composing shapes. The `RSMetricsProvider` class is the root class of the supported strategy and may be set in a `RSLabel` using the `metricsProvider:` method.

Each label is associated with a font. It may be set using the `font:` method. Consider the following example (see Figure 7-6):

```
c := RSCanvas new.
font := LogicalFont familyName: 'Source Code Pro' pointSize: 20.
words := String loremIpsum splitOn: ' '.
c addAll: (words collect: [ :word |
 RSLabel new
  font: font;
```

```
  text: word;
  yourself ]).
RSFlowLayout new gapSize: 10; on: c shapes.
c shapes @ RSHighlightable red.
c shapes @ RSDraggable.
c @ RSCanvasController.
c open
```

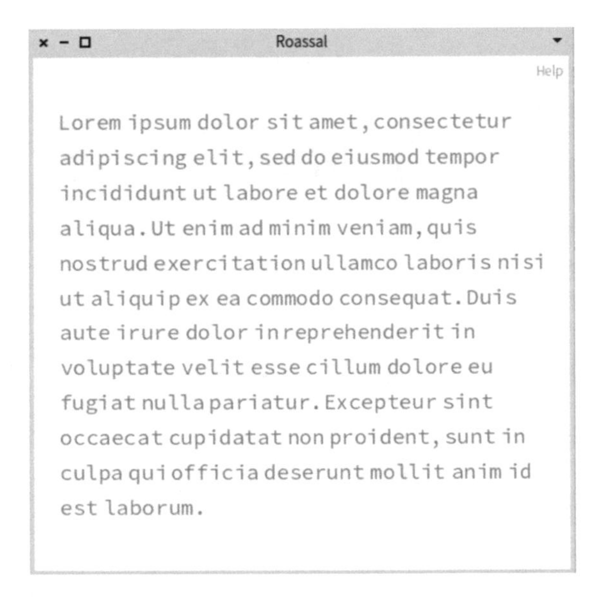

Figure 7-6. *Setting a font*

Obviously, this script will work as expected if the Source Code Pro font is locally installed on your operating system. The list of available font names may be obtained by inspecting the following expression:

```
FreeTypeFontProvider current updateFromSystem families
```

Some font provides attributes can be used in a label. Consider the following example (see Figure 7-7):

```
styles := #(#italic #bold #normal #struckOut #underline).

c := RSCanvas new.
styles do: [ :aStyle |
    label := RSLabel text: aStyle asString.
```

```
    label perform: aStyle.
    c add: label ].

RSVerticalLineLayout on: c shapes.
c @ RSCanvasController.
c open
```

Figure 7-7. *Font attributes*

Polygon

Polygons can be defined using the RSPolygon class. A polygon is defined as a set of controlled points. Consider the following example (see Figure 7-8):

```
c := RSCanvas new.

polygon := RSPolygon new
            points: { 0 @ -50 . 50 @ 0 . -50 @ 0 };
            color: 'FFAE0B'.
polygon cornerRadii: 5.
polygon @ RSDraggable.
```

```
polygon2 := RSPolygon new
            points: { 0 @ -50 . 50 @ 0 . -50 @ 0 };
            color: Color red translucent.
polygon2 @ RSDraggable.

polygon2 rotateByDegrees: 90.
polygon2 translateBy: 0 @ -50.

c add: polygon.
c add: polygon2.

c zoomToFit.
c open
```

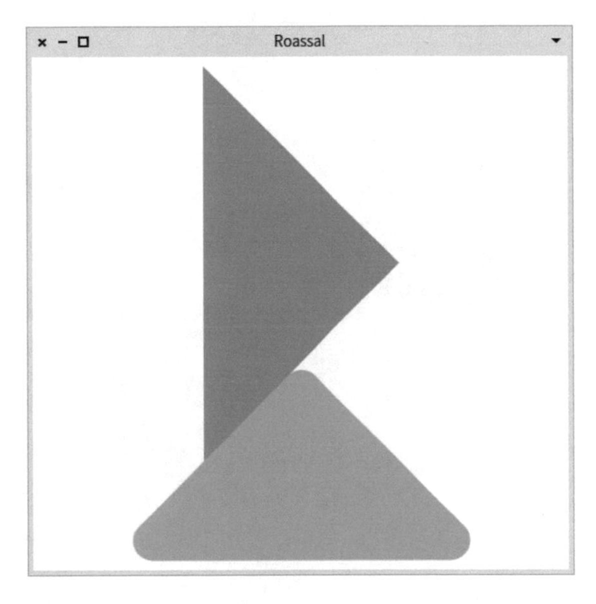

Figure 7-8. *Example of two polygons*

You can round a polygon's corners using `cornerRadii:`, which expects a positive integer as an argument.

SVG Path

SVG is an expressive format to define a complex visual element. Roassal offers the RSSVGPath class to represent the SVG path as a shape. Consider the following example (see Figure 7-9):

```
c := RSCanvas new.
svgPath := 'M 10 80 C 40 10, 65 10, 95 80 S 150 150, 180 80'.
svg := RSSVGPath new svgPath: svgPath.
c add: svg.
c @ RSCanvasController.
c open
```

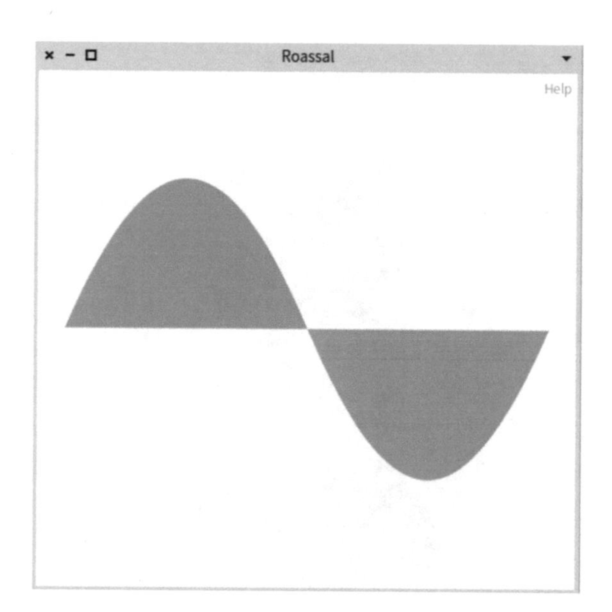

Figure 7-9. *Example of a SVG path*

A SVG shape requires a string describing its path, itself encoded into some commands. For example, the token M is the command *moveto*, C is *curveto*, and S is *smooth curveto*. A description of these tokens can easily be found online. w3schools.com is a reliable source of documentation.

Numerous online services offer SVG paths ready to be consumed. Most search engines will indicate online resources to obtain paths. SVG paths can be used to define complex transformable icons. A slightly more complex example is shown in the following code (see Figure 7-10):

```
c := RSCanvas new.
svgPaths :=
    #('M17.46,22H6.54a4.55,4.55,0,0,1-3.42-7.54L9,7.75V4a1,1,0,0,1,2,0V
    8.12a1,1,0,0,1-.25.66l-6.12,7A2.54,2.54,0,0,0,6.54,20H17.46a2.54,
    2.54,0,0,0,1.91-4.22l-6.12-7A1,1,0,0,1,13,8.12V6.5a1,1,0,0,1,2,
    0V7.75l5.88,6.71A4.55,4.55,0,0,1,17.46,22Z'
    'M15,4.12H9a1,1,0,0,1,0-2h6a1,1,0,0,1,0,2Z'
    'M19,15H10a1,1,0,0,1,0-2h9a1,1,0,0,1,0,2Z'
    'M7,18a1,1,0,0,1-1-1,1,1,0,0,1,.08-.38.93.93,0,0,1,.21-.33,1,1,0,0,
    1,1.42,0,1,1,0,0,1,.21.33A.84.84,0,0,1,8,17a1,1,0,0,1-1,1Z'
    'M11,21a1,1,0,0,1-.38-.08.93.93,0,0,1-.33-.21,1,1,0,0,1,
    0-1.42.93.93,0,0,1,.33-.21,1,1,0,0,1,1.09.21A1,1,0,0,1,11,21Z'
    'M15,18l-.19,0a.6.6,0,0,1-.19-.06.76.76,0,0,
    1-.18-.09l-.15-.12A1.05,1.05,0,0,1,14,17a1,1,0,0,1,.08-.38.93.93,0,0,1,
    .21-.33,1.58,1.58,0,0,1,.15-.12.76.76,0,0,1,.18-.09.6.6,0,0,1,
    .19-.06,1,1,0,0,1,.9.27.93.93,0,0,1,.21.33A1,1,0,0,1,16,17a1,1,0,0,
    1-1,1Z'
    'M12,12a1.05,1.05,0,0,1-.71-.29,1.15,1.15,0,0,1-.21-.33.94.94,0,0,
    1,0-.76,1,1,0,0,1,.21-.33A1,1,0,0,1,12.2,
    10l.18.06.18.09.15.12a1.15,1.15,0,0,1,.21.33A1,1,0,0,1,13,11a1,1,
    0,0,1-1,1Z'
    ).

25 timesRepeat: [
    aRandomColor := Color random translucent.
    svg := svgPaths collect: [ :path | RSSVGPath new svgPath: path ] as:
    RSGroup.
    svg color: aRandomColor.
    shape := svg asShape.
    c add: shape.
].
```

```
RSGridLayout new lineItemsCount: 5; on: c shapes.
c @ RSCanvasController.
c open
```

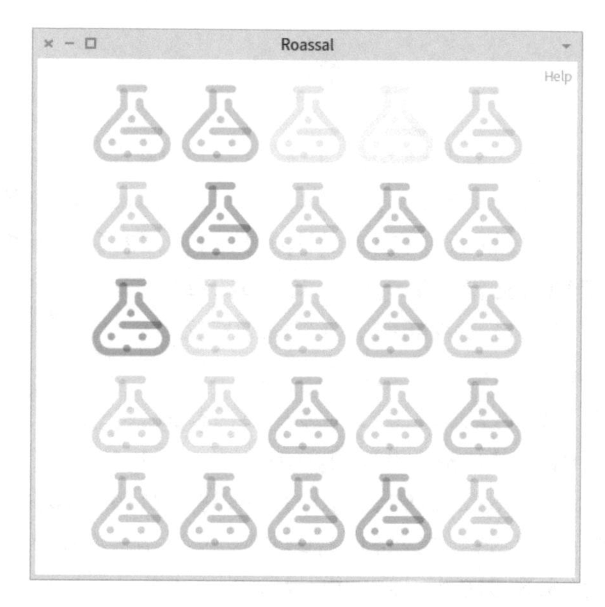

Figure 7-10. *Example of SVG-based icons*

These SVG paths are from `svgrepo.com` and are distributed under the public domain license.

Common Features

All the shapes described so far inherit from the `RSBoundingShape` class. Therefore, a number of common features are supported for all these shapes.

Color may be set using `color:`, which accepts either a color object (e.g., `Color blue` or `Color r: 0.5 g: 0.3 b: 0.2`) or a symbol (e.g., `#red`, `#cyan`). If a symbol is provided, it is executed on the `Color` class. Methods defined on the class side of `Color` represent valid symbols to be provided to `color:`.

As mentioned, each shape has a position, which can be modified at will. A shape can be translated using:

- `translateBy: aDeltaPoint`, which moves a shape by a given step expressed as a point.
- `translateTo: aPoint`, which moves a shape to a new position.

These two methods move a shape according to its center. A shape may be translated by considering a reference point as a corner. The following methods may be executed: `translateTopLeftTo:`, `translateTopRightTo:`, `translateBottomRightTo:`, and `translateBottomLeftTo:`. For example, `translateTopLeftTo: aPoint` moves the top-left corner of a shape to a particular location. These methods help you easily fine-tune the shapes' positions. As an illustration, the following script locates a label on the top-left corner of the window:

```
c := RSCanvas new.
lbl := RSLabel text: 'Top left corner'.
c add: lbl.
lbl setAsFixed.
lbl translateTopLeftTo: 0 @ 0.
c open
```

Each shape contains a transformation matrix that is useful to express geometrical operations. In particular, a shape offers the `scaleBy:` and `rotateByDegrees:` methods. Here is an example of rotating a shape (see Figure 7-11):

```
c := RSCanvas new.

(0 to: 90 count: 10) do: [ :rotation |
    lbl := RSLabel text: 'Hello world'.
    lbl color: Color gray translucent.
    lbl rotateByDegrees: rotation.
    lbl @ RSHighlightable red.
    lbl translateTopLeftTo: 0 @ 0.
    c add: lbl ].
c @ RSCanvasController.
c open
```

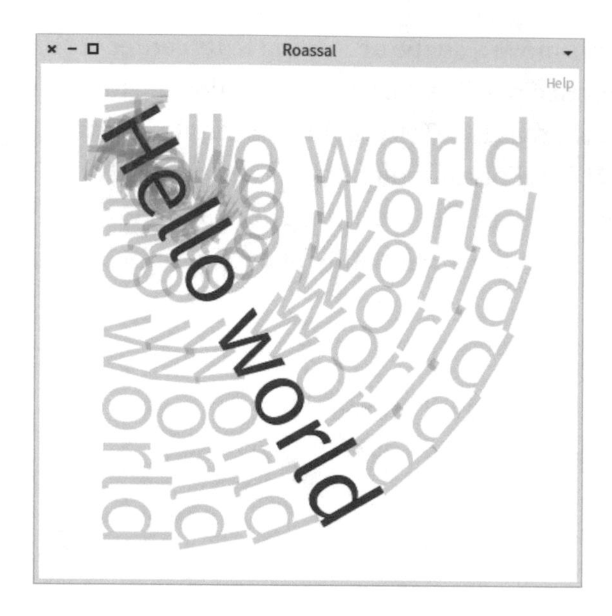

Figure 7-11. *Rotating a shape*

A border can be set on a shape by using `borderColor:`, which accepts a color (object or symbol) as an argument. Consider the example in the following code (see Figure 7-12):

```
c := RSCanvas new.
r := Random seed: 42.
40 timesRepeat: [
    box := RSBox new
            width: (r nextInteger: 80);
            height: (r nextInteger: 80);
            color: Color gray;
            cornerRadius: 10.
    box borderColor: #black.
    box color: Color random translucent.
    box translateTo: (r nextInteger: 200) @ (r nextInteger: 200).
    c add: box.
].
c @ RSCanvasController.
c open
```

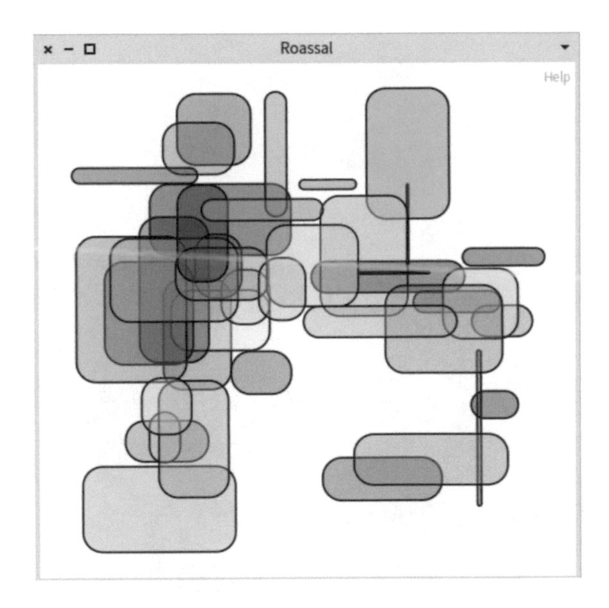

Figure 7-12. *Rounded boxes with a thin black border*

Borders are modeled with the RSBorder class, which offers many possibilities. For example, a dashed pattern can be provided using dashArray:. Consider the following example, which sets an animation on the dash (see Figure 7-13):

```
c := RSCanvas new.

b := RSBorder new color: Color blue.
b dashArray: #(5 1 5).

(30 to: 60 by: 5) do: [ :nb |
 box := RSBox new size: nb; cornerRadius: 10.
 ellipse := RSEllipse new width: nb; height: nb + 10.
 box border: b.
 ellipse border: b.
 c add: box; add: ellipse ].

RSFlowLayout on: c shapes.
c @ RSCanvasController.
```

```
c newAnimation
 from: 0;
 to: 40;
 on: b set: #dashOffset:.
c open
```

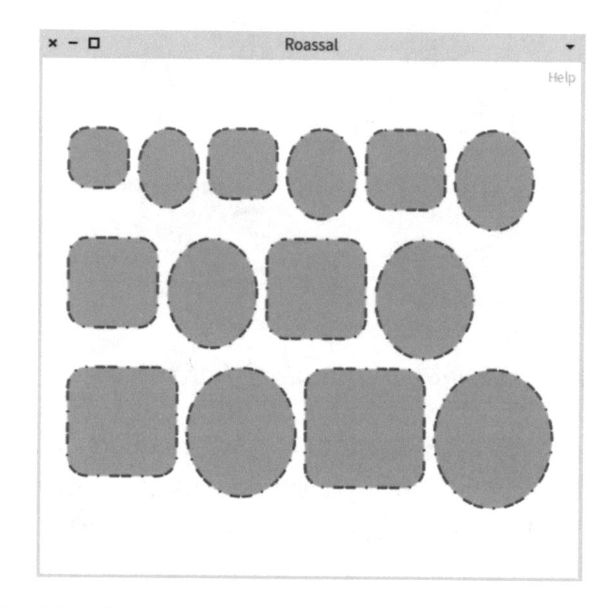

Figure 7-13. *Dashed borders*

I encourage you to browse the definition of the `RSBorder` class to see an exhaustive list of how borders are handled.

Model

A shape can have a model object that it is intended to represent. All the shapes defined in this chapter do not have a model object. Setting a model in a shape is useful to define lines, use a normalizer, and set interactions. Having shapes that support a model object is a significant advantage that Roassal has over other visualization engines.

A model is simply set on a shape using `model:`. For example, you can define a group of circles, each representing a number, as follows:

```
c := RSCanvas new.
numbers := 1 to: 9.
```

```
numbers do: [ :nb |
    circle := RSCircle new.
    circle model: nb.
    circle @ RSPopup.
    c add: circle ].
RSGridLayout on: c shapes.
c zoomToFit.
c open
```

The script creates nine circles in a canvas. Each circle has a number, and the represented numbers range from 1 to 9. Each circle has a popup interaction, which makes a little window appear when the mouse is above a circle. The popup interaction obtains a textual representation of each model object to define the small window's content.

This revision of the previous example uses a normalizer to assign a particular size and color to each circle (see Figure 7-14):

```
c := RSCanvas new.
numbers := 1 to: 9.
numbers do: [ :nb |
    circle := RSCircle new.
    circle model: nb.
    circle @ RSPopup.
    c add: circle ].

RSNormalizer size
    shapes: c nodes;
    from: 5; to: 20;
    normalize: #yourself.

RSNormalizer color
    shapes: c nodes;
    from: Color gray; to: Color red;
    normalize: [ :aNumber | aNumber raisedTo: 3 ].

RSFlowLayout on: c shapes.
c zoomToFit.
c open
```

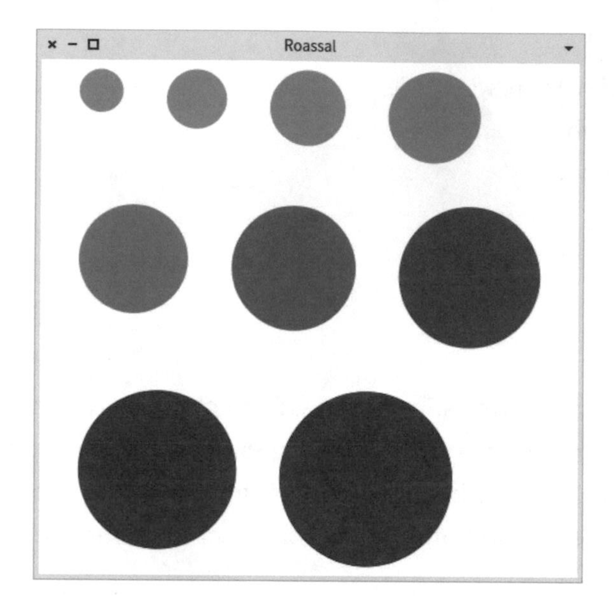

Figure 7-14. *Normalizing the size and color using models*

Another significant benefit of using a model object is to enjoy the support of the Pharo's Inspector. Chapter 13 is dedicated to that topic.

Line

Although apparently simple, representing lines remains a complex task. As any other shape, a line has many visual attributes (e.g., width, color, style, and shapes as extremities). Difficulties in handling lines include the way they are built and how to react to user interactions. In its simplest form, a line may be set between two points. However, the way a line has to be configured becomes more complex when a line connect two shapes. For example, the line needs to be adapted when one of its extremities is translated.

Lines are defined by `RSAbstractLine` and its subclasses:

- `RSLine` describes a direct line between two points or shapes.
- `RSPolyline` represents a line with multiple control points.
- `RSBezier` represents a Bezier line with two, three, or four control points.
- `RSArrowedLine` models a straight arrowed line between two points.

RSLine is the most commonly employed line. Consider the following example:

```
c := RSCanvas new.
line := RSLine new.
line from: 30 @ 40.
line to: 150 @ 120.
c add: line.
c open
```

In this example, the two extremities of the line are specified as points. A line is drawn from the point (30, 40) to (150, 120). A line can also connect two shapes. In this case, a shape can be provided when invoking from: and to:. Consider the following example:

```
c := RSCanvas new.

box := RSBox new.
circle := RSCircle new.

c add: box; add: circle.
box @ RSDraggable.
circle @ RSDraggable.

circle translateTo: 50 @ 40.

line := RSLine new.
line from: box.
line to: circle.
c add: line.

c open
```

The two shapes—box and circle—are draggable. Moving one of these leaves the line properly attached to the shape.

Line Attach Point

The junction between a line and a shape is driven by a particular object, called an attach point. An attach point can be set in a line by simply sending `attachPoint:` to it with the attach point as the parameter. Consider the example in the following code (see Figure 7-15):

```
c := RSCanvas new.

box := RSBox new.
circle := RSCircle new.

c add: box; add: circle.
box @ RSDraggable.
circle @ RSDraggable.

"We make the box and circle translucent"
{ box . circle } asGroup translucent.

circle translateTo: 50 @ 40.

line := RSLine new.
line color: Color red.
line from: box.
line to: circle.
line attachPoint: RSBorderAttachPoint new.
c add: line.

c zoomToFit.
c open
```

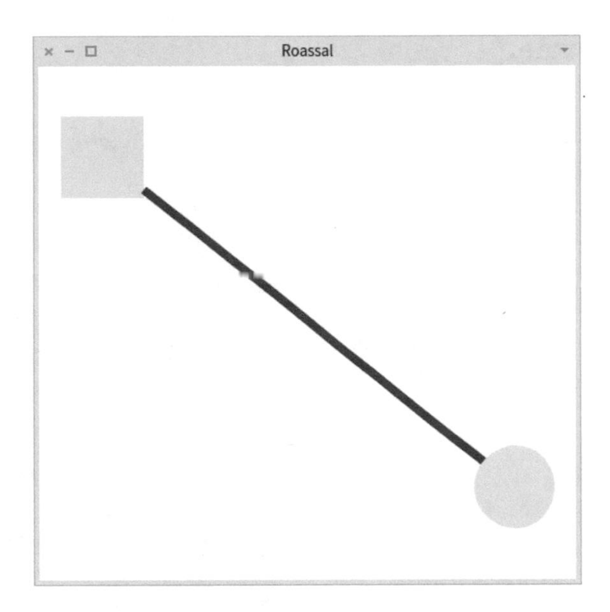

Figure 7-15. *Using a border attach point*

The previous example uses a border attach point as a way to make the extremities stick to the border of those shapes. The `RSAttachPoint` class is the root of the attach point class hierarchy. It may happen that some layout requires particular attach points. For example, a vertical tree layout may be combined with a vertical attach point.

The `RSAbstractLine` class provides useful shortcut methods, including `withBorderAttachPoint`, `withCenteredAttachPoint`, `withHorizontalAttachPoint`, and `withVerticalAttachPoint`. In the previous script, you can replace the `line attachPoint: RSBorderAttachPoint new` expression with `line withBorderAttachPoint`.

Line Marker

A marker is a decoration that can be set on a line. A line may have zero, one, or more markers, located at a position anywhere between the two extremities. Any shape could serve as a marker by simply sending `asMarker` to it. Consider the following example (see Figure 7-16):

```
c := RSCanvas new.

box := RSBox new.
circle := RSCircle new.
```

```
c add: box; add: circle.
box @ RSDraggable.
circle @ RSDraggable.

circle translateTo: 50 @ 40.

line := RSLine new.
line withBorderAttachPoint.
line marker: (RSCircle new size: 5; color: #red) asMarker.
line from: box.
line to: circle.
c add: line.
c zoomToFit.
c open
```

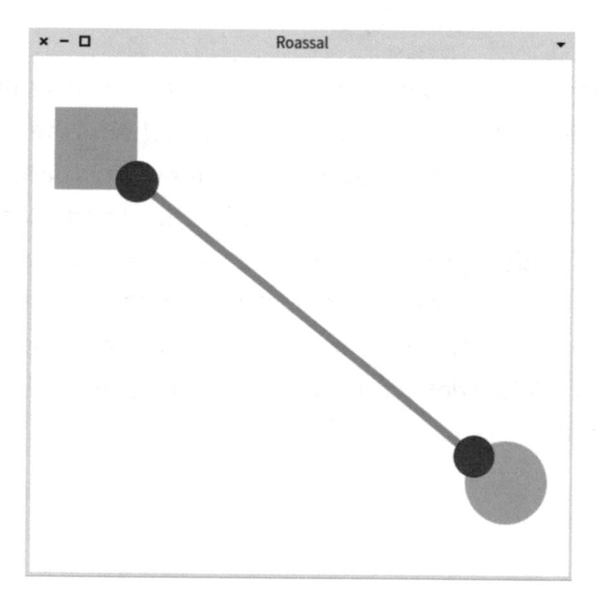

Figure 7-16. *Line with a marker*

The RSAbstractLine class offers an API to define and configure markers. A shape is converted to a marker by simply sending asMarker to it, returning an instance of the class RSMarker. A marker is added to a line shape using one of these messages:

- markerStart: To set the provided marker at the origin of the line, i.e., where the line starts.

- `markerEnd`: To set the marker at the target, i.e., where the line ends.
- `marker`: To set the marker at both extremities of the line.

To see the effect of positioning the marker, consider the following code snippet (see Figure 7-17):

```
c := RSCanvas new.

markerShape := RSPolygon new
        privatePoints: { -5@9 . 0@0 . 5@9 . 0@0 };
        border: (RSBorder new width: 1);
        asMarker.

labelStart := RSLabel text: 'start'.
labelEnd := RSLabel text: 'end'.
labelStart @ RSDraggable.
labelEnd @ RSDraggable.
c add: labelStart.
c add: labelEnd.

labelEnd translateBy: 80 @ 60.

line := RSLine new.
line from: labelStart.
line to: labelEnd.
line withBorderAttachPoint.
line markerEnd: markerShape.

c add: line.
c zoomToFit.
c open
```

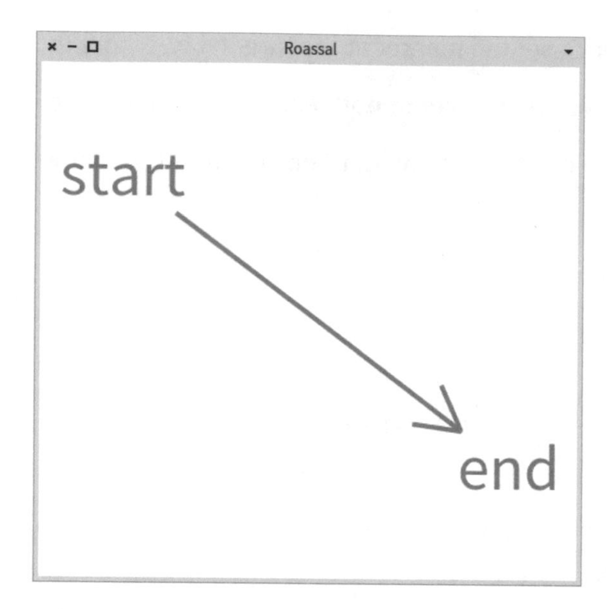

Figure 7-17. *Line with a marker*

Since arrowed line are commonly used, a dedicated class, called `RSArrowedLine`, is provided for that purpose. The previous script can be rewritten using an arrowed line, as follows:

```
c := RSCanvas new.

labelStart := RSLabel text: 'start'.
labelEnd := RSLabel text: 'end'.
labelStart @ RSDraggable.
labelEnd @ RSDraggable.
c add: labelStart.
c add: labelEnd.

labelEnd translateBy: 80 @ 60.

line := RSArrowedLine new.
line from: labelStart.
line to: labelEnd.
line withBorderAttachPoint.

c add: line.
c zoomToFit.
c open
```

The RSArrowedLine class can be used to produce arrowed lines without explicitly defining markers. In addition to configuring the visual shape of a marker, the position of a marker on the line may be set using two methods:

- offset: Sets a fixed number of pixels between the extremity and the marker.
- offsetRatio: Sets a ratio, between 0.0 and 1.0, to determine the position of the marker.

Moving the marker away from the extremity is very useful to avoid arrow cluttering (see Figure 7-18):

```
c := RSCanvas new.

markerShape := RSPolygon new
        privatePoints: { -3 @ 9 . 0 @ 0 . 3 @ 9 . 0 @ 0 };
        border: (RSBorder new width: 1);
        asMarker.
markerShape offsetRatio: 0.3.

target := RSLabel text: 'target'.
c add: target.

starts := RSCircle models: (1 to: 20).
starts @ RSDraggable.
c addAll: starts.

starts do: [ :startShape |
    line := RSLine new.
    line from: startShape.
    line to: target.
    line withBorderAttachPoint.
    line markerEnd: markerShape.
    c add: line.
].

RSForceBasedLayout new charge: -300; on: c nodes.

c zoomToFit.
c open
```

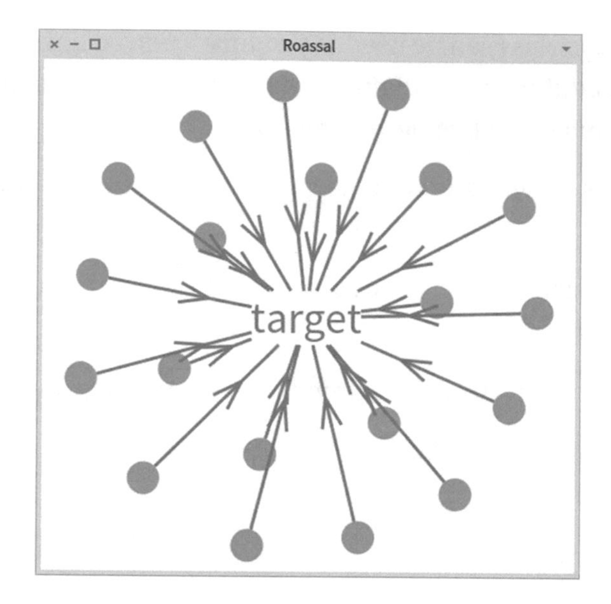

Figure 7-18. *Offset on arrow markers*

Removing the `markerShape offsetRatio: 0.3.` line has the effect of having all the markers on the border of the label target. All the markers superimpose each other, which defeats the purpose of having markers. Having distance between the marker and the target is useful to avoid clutter.

Line with Control Points

The `RSLine` and `RSArrowedLine` classes represent a straight line between two extremities. `RSBezier` and `RSPolyline` are two kinds of lines that accept control points. Consider the following example (see Figure 7-19):

```
canvas := RSCanvas new.
canvas add: (RSBezier new
            color: Color red;
            controlPoints: { (0 @ 0). (100 @ 100). (200 @ 0).
            (300 @ 100)}).
canvas open
```

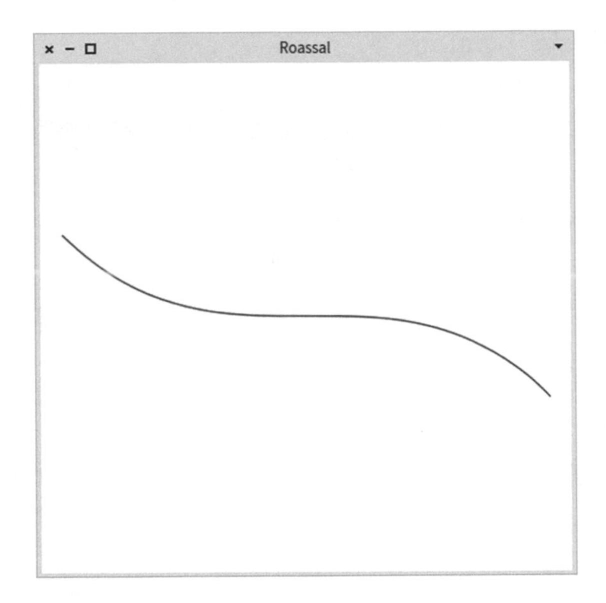

Figure 7-19. *A simple Bezier line*

In the previous example, the control points are explicitly provided and remain static. If the extremities of a Bezier line can be dragged, the control points cannot be static anymore and must be determined. For that purpose, Roassal offers a class hierarchy for controlled points and its root class is `RSAbstractCPController`. Consider the following example (see Figure 7-20):

```
canvas := RSCanvas new.

box1 := RSBox new color: Color blue.
box2 := RSBox new color: Color red.
box2 translateTo: 100 @ -200.

box1 @ RSDraggable.
box2 @ RSDraggable.
canvas add: box1; add: box2.

bezierLine := RSBezier new
    withVerticalAttachPoint;
    from: box1;
    to: box2;
    controlPointsController: (
            RSBlockCPController new
```

```
                block: [ :aLine |
                    | mid |
                       mid := (box1 position + box2 position) / 2.
                    {(box1 position) .
                    (box1 position x @ mid y) .
                    (box2 position x @ mid y) .
                    (box2 position)} ];
                yourself);
  yourself.
canvas add: bezierLine.
canvas zoomToFit.
canvas open
```

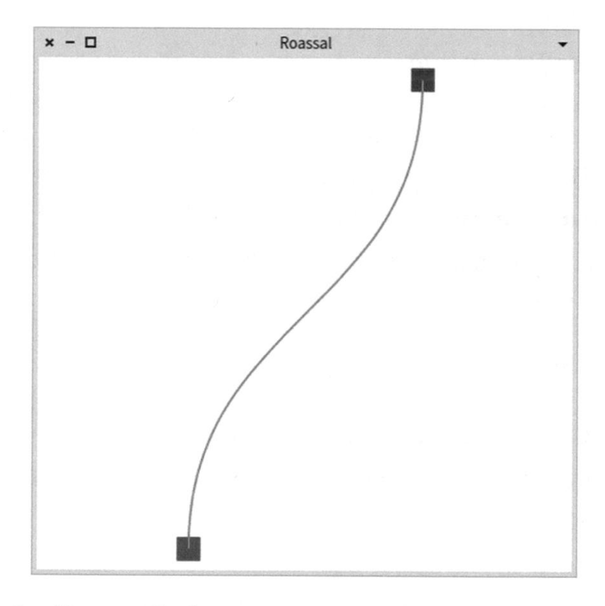

Figure 7-20. *A Bezier line with dynamic controlled points*

Control points are computed at each translation of the extremities when using `RSBlockCPController`. The block takes a line as an argument and has to return an array of three or four points. A number of controllers are provided. For example, the previous script can be rewritten:

```
canvas := RSCanvas new.
```

```
box1 := RSBox new color: Color blue.
box2 := RSBox new color: Color red.
box2 translateTo: 100 @ -200.

box1 @ RSDraggable.
box2 @ RSDraggable.
canvas add: box1; add: box2.

bezierLine := RSBezier new
    withVerticalAttachPoint;
    from: box2;
    to: box1;
    controlPointsController: RSVerticalCPAPController new;
  yourself.
canvas add: bezierLine.
canvas zoomToFit.
canvas open
```

The `RSVerticalCPAPController` controller replaces the control points that are manually specified. Similarly, `RSHorizontalCPAPController` defines control points that are adequate for an horizontal alignment.

What Have You Learned in This Chapter?

This chapter presents the essential shapes provided by Roassal. It focuses on the following:

- Simple shapes, including boxes, circles, ellipses, and labels
- The benefits of having a model behind a shape
- A line may be set between two points or two shapes
- Bezier lines and polylines can have controlled points, statically or dynamically computed

The coming chapters will build on top of the notions presented in this chapter.

CHAPTER 8

Line Builder

This chapter describes an expressive way to build lines in Roassal. I'll start by illustrating the difficulties in defining lines. Subsequently, I'll demonstrate how Line Builder is used to address these difficulties.

All the code provided in this chapter is available at `https://github.com/bergel/AgileVisualizationAPressCode/blob/main/02-04-LineBuilder.txt`.

Difficulties with Build Lines

A visualization typically represents a particular set of elements and logical structure is best expressed using lines. Manually maintaining the association between the represented elements and the visual representation can be very cumbersome. Consider the following example, borrowed from the field of software visualization. In Pharo, a software component is typically expressed in term of classes, which are possibly linked to each other using inheritance. Visually representing the inheritance of a class hierarchy is a powerful mechanism that the UML class diagram is based on. Consider the following code snippet:

```
classes := Collection withAllSubclasses.

c := RSCanvas new.

"Build nodes"
boxes := RSCircle models: classes.
c addAll: boxes.
RSFlowLayout on: c shapes.

"Connect nodes"
classes do: [ :cls |
    fromBox := boxes shapeFromModel: cls superclass.
    toBox := boxes shapeFromModel: cls.
```


A. Bergel, *Agile Visualization with Pharo*, https://doi.org/10.1007/978-1-4842-7161-2_8

```
    fromBox notNil ifTrue: [
        line := RSLine new.
        line from: fromBox.
        line to: toBox.
        c add: line ].
    ].

RSTreeLayout on: c nodes.
c @ RSCanvasController.
c open
```

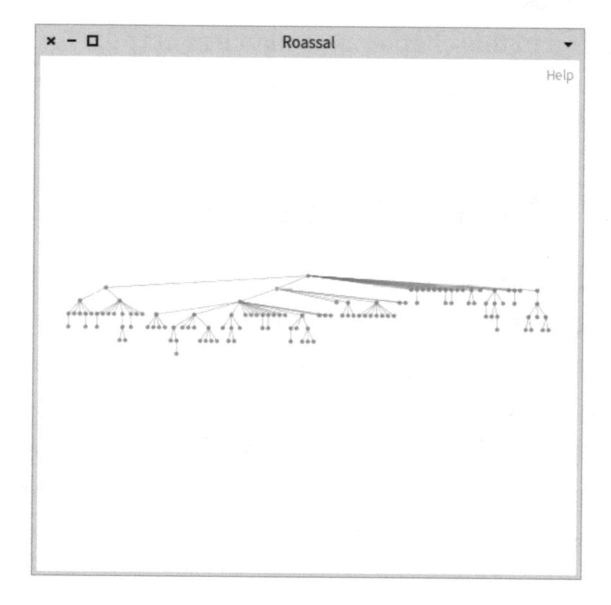

Figure 8-1. *Joining each class to its superclass*

The way that nodes are connected is rather complex. First, an iteration is explicitly performed to retrieve the two extremities using `shapeFromModel:`. However, the box representing a superclass might not exist, as is the case of the `Collection` class, whose superclass is `Object`, and is not contained in the `classes` variable.

Using a Line Builder

The RSLineBuilder class greatly simplifies the creation of lines. Consider this improved version of the code given previously:

```
classes := Collection withAllSubclasses.

c := RSCanvas new.

boxes := RSCircle models: classes.
c addAll: boxes.
RSFlowLayout on: c shapes.

lb := RSLineBuilder line.
lb shapes: boxes.
lb connectFrom: #superclass.

RSTreeLayout on: c nodes.
c @ RSCanvasController.
c open
```

The result of this improved version is identical to the original. The benefits of using the Line Builder are obvious. There is:

- No need to "manually" iterate over a collection to retrieve the shape from a given object
- No need to check if a shape is missing

Furthermore, Line Builder offers numerous methods to easily create lines. For example:

```
classes := Collection withAllSubclasses.

c := RSCanvas new.

boxes := RSCircle models: classes.
c addAll: boxes.
RSFlowLayout on: c shapes.

lb := RSLineBuilder orthoVertical.
lb withVerticalAttachPoint.
```

```
lb capRound.
lb shapes: boxes.
lb connectFrom: #superclass.

RSTreeLayout on: c nodes.

c @ RSCanvasController.
c open
```

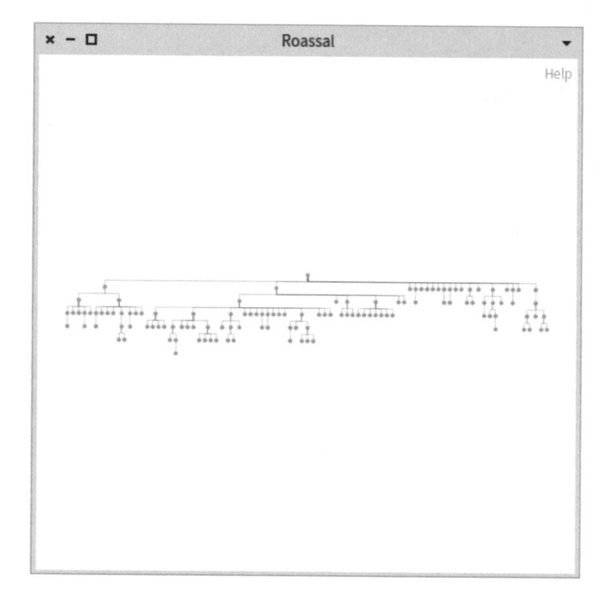

Figure 8-2. *Using a Line Builder*

Line Builder is used to produce ortho-vertical lines instead of straight lines. By default, a line connect shapes from their centers. This default behavior can be overridden using `withVerticalAttachPoint` in this example. Corners of the ortho-vertical lines are rounded using `capRound` (this effect is apparent when zooming in).

The builder looks for extremities in the shapes, which are provided using `shapes:`. Each line produced from a shape has the "from" extremity referring to the superclass.

Using Associations

In Pharo, the method called `->` creates an association, an instance to the `Association` class. An association links a key to a value. Due to its syntactic conciseness, associations are handy ways to specify connections between elements.

Line Builder can use associations to build lines. Consider this script (see Figure 8-3):

```
b := RSCanvas  new.
b addAll: (RSCircle models: (1 to: 4)).
b shapes @ RSDraggable @ RSPopup.

RSLineBuilder line
    color: Color red translucent;
    canvas: b;
    withBorderAttachPoint;
    useAssociations: { 1 -> 2 . 2 -> 3 . 4 -> 1 . 3 -> 4 }.

RSGridLayout new gapSize: 30; on: b nodes.

b @ RSCanvasController.
b open
```

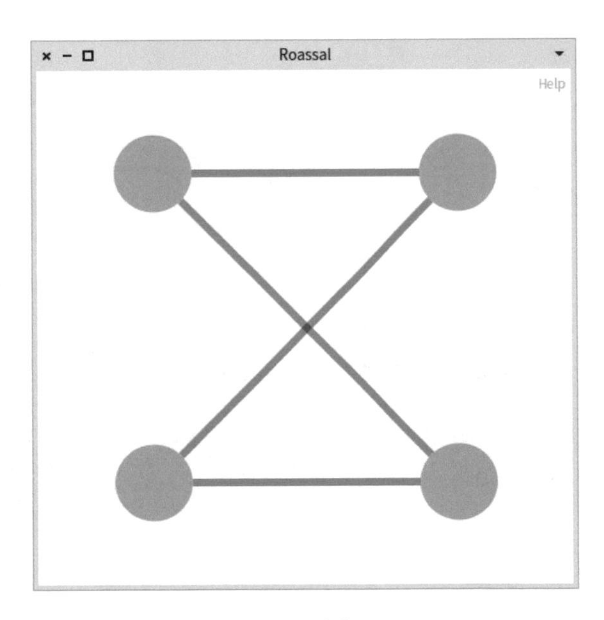

Figure 8-3. *Graph building with Line Builder*

Graph Visualization

Line Builder significantly reduces the effort needed to specify connections, as commonly used to represent graphs. Consider the following code snippet (see Figure 8-4):

```
numberOfNodes := 25.
numberOfLines := 90.
r := Random seed: 42.
graph := Dictionary new.

1 to: numberOfNodes do: [ :aNode |
    graph at: aNode put: Set new ].

numberOfLines timesRepeat: [
    fromNode := r nextInteger: numberOfNodes.
    toNode := r nextInteger: numberOfNodes.
    (graph at: fromNode) add: toNode ].

canvas := RSCanvas new.
nodes := RSLabel models: (1 to: numberOfNodes).
nodes color: #red.
nodes @ RSDraggable @ RSPopup.
canvas addAll: nodes.

lb := RSLineBuilder line.
lb canvas: canvas.
lb withBorderAttachPoint.
lb makeBidirectional.
lb moveBehind.
lb objects: (1 to: numberOfNodes).
lb connectToAll: [ :aNumber | graph at: aNumber ].

RSForceBasedLayout new charge: -300; on: nodes.
canvas @ RSCanvasController.
canvas open
```

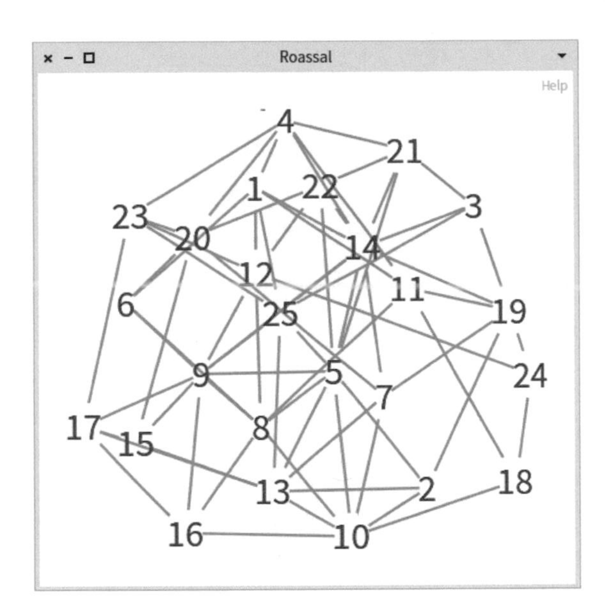

Figure 8-4. *Graph visualization*

The beginning of the script generates a graph made of `numberOfNodes` nodes and `numberOfLines` lines. The `graph` variable is a dictionary that contains connections. The Line Builder object, `lb`, is in charge of drawing edges between nodes. The builder moves the lines behind the node using `moveBehind`. This is useful since having lines in front may hide the nodes. A border attach point is set, and the lines are anchored on the border of each label, the one closest to the other extremity. Furthermore, lines are bidirectional to avoid redundant lines (e.g., only one line from `1- > 2` and `2- > 1` is constructed). This is useful since the graph is randomly generated.

This example illustrates the usefulness of Line Builder. The same example without Line Builder involves a significant amount of code.

What Have You Learned in This Chapter?

Line Builder is frequently used due to its expressiveness. Line Builder is an important asset of Roassal, and I recommend carefully studying it. In particular, Line Builder aims at:

- Simplifying the construction of many lines.
- Providing utility methods to control the placement and generation of lines.

CHAPTER 9

Shape Composition

The previous chapters presented primitive shapes that are offered by Roassal. An important asset of Roassal is the composition of visual elements. This chapter reviews the mechanism to build composed shapes from primitives shapes.

All the code provided in this chapter is available at https://github.com/bergel/AgileVisualizationAPressCode/blob/main/02-05-ShapeComposition.txt.

Composite Shapes

Shapes can be composed in order to build sophisticated visual components. The RSGroup class is a collection of shapes and offers various convenient methods to perform operations on a set of shapes. A simple way to create a composite shape is to send asShape to an RSGroup. This message is used to convert a group of shapes into a single composed shape. Consider the following example (see Figure 9-1):

```
canvas := RSCanvas new.

1 to: 10 do: [ :i |
    g := RSGroup new.
    i timesRepeat: [ g add: RSCircle new ].
    RSFlowLayout on: g.
    canvas add: g asShape.
].

RSGridLayout new gapSize: 50; on: canvas shapes.
canvas shapes @ RSDraggable @ RSHighlightable red.
canvas zoomToFit.
canvas open
```


A. Bergel, *Agile Visualization with Pharo*, https://doi.org/10.1007/978-1-4842-7161-2_9

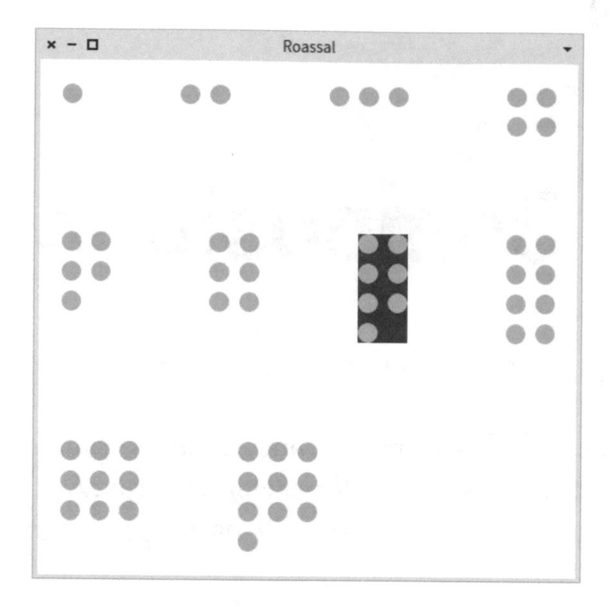

***Figure 9-1.** Simple example of composed shapes*

A layout and interactions can be applied to the added shapes, pretty much the same way that you do with non-composite shapes. In thc loop, you create a group g to which you add some shapes. A flow layout is performed on the shapes to be composed using `RSFlowLayout on: g`. The group is converted into a composite shape with `g asShape`. In total, ten shapes, all composite, are directly added to the canvas. The grid layout operates on these composite shapes. Each of these can be highlighted by moving the mouse above them.

A shape can also have a border, simply by sending `borderColor:`. Often, padding is necessary to set the distance between the nested shapes and the border. The padding is set using `padding:` and expects a number as the argument. Consider the revised version of the script (see Figure 9-2):

```
canvas := RSCanvas new.

1 to: 10 do: [ :i |
    g := RSGroup new.
    i timesRepeat: [ g add: RSCircle new ].
    RSFlowLayout on: g.
    compositeShape := g asShape.
    compositeShape borderColor: Color black.
```

```
    compositeShape padding: 5.
    canvas add: compositeShape.
].

RSGridLayout new gapSize: 50; on: canvas shapes.
canvas shapes @ RSDraggable @ RSHighlightable red.
canvas zoomToFit.
canvas open
```

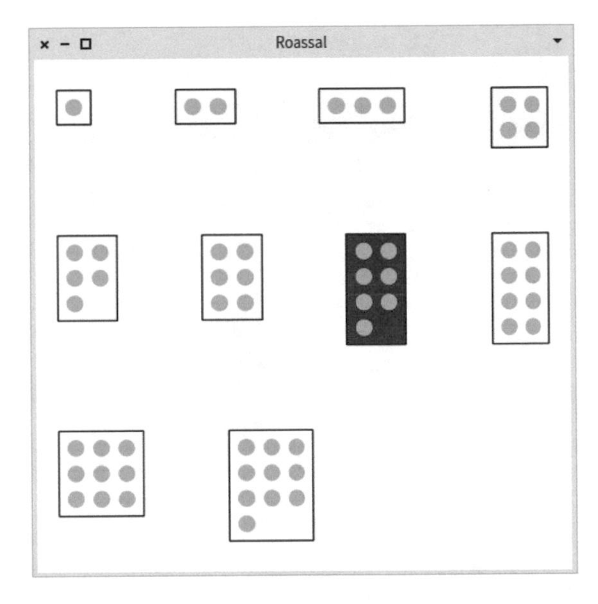

***Figure 9-2.** Setting a border with a padding to the composed shapes*

Labels may be part of the composition. This is often useful for naming a composite shape. Consider this third revision of the script (see Figure 9-3):

```
canvas := RSCanvas new.

1 to: 10 do: [ :i |
    g := RSGroup new.
    i timesRepeat: [ g add: RSCircle new ].
    RSFlowLayout on: g.

    title := RSLabel text: i.
    RSVerticalLineLayout new alignCenter; on: { title . g }.
    g add: title.
```

```
    compositeShape := g asShape.
    compositeShape borderColor: Color black.
    compositeShape padding: 5.
    canvas add: compositeShape.
].

RSGridLayout new gapSize: 50; on: canvas shapes.
canvas shapes @ RSDraggable @ RSHighlightable red.
canvas zoomToFit.
canvas open
```

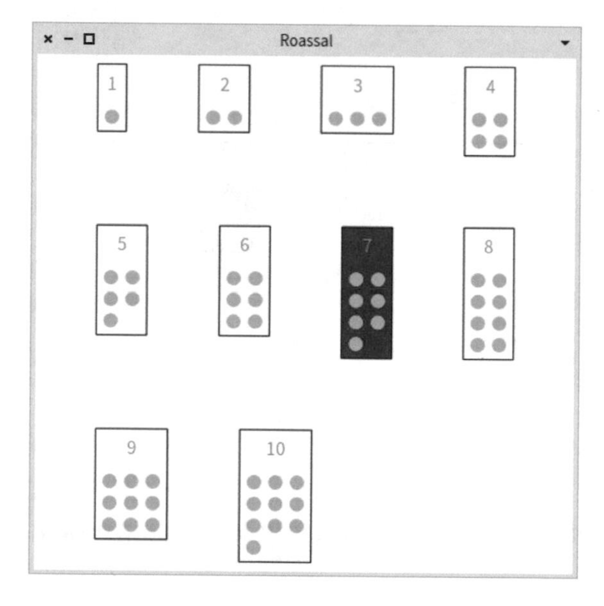

Figure 9-3. *Adding a title to the composed shape*

In the script, a title is created, and a vertical line layout is performed on two elements: `title` and g. After the layout, the title is added to the group. This example illustrates the fact that shapes can go through a complex layout before being transformed into a composed shape.

Lines can also be added to a composite shape. Consider the following code (see Figure 9-4):

```
canvas := RSCanvas new.

1 to: 100 by: 10 do: [ :i |
    g := RSCircle models: (1 to: i) forEach: [ :s :nb | s color: Color
    random translucent ].

    lb := RSLineBuilder line.
    lb canvas: g.
    lb shapes: g.
    lb connectFrom: [ :nb | nb // 2 ].

    RSRadialTreeLayout on: g nodes.

    title := RSLabel new fontSize: 20; text: i.
    RSVerticalLineLayout new alignCenter; on: { title . g }.
    g add: title.

    compositeShape := g asShape.
    compositeShape borderColor: Color black.
    compositeShape padding: 5.
    canvas add: compositeShape.
].

RSGridLayout new gapSize: 50; on: canvas shapes.
canvas shapes @ RSDraggable @ RSHighlightable red.
canvas zoomToFit.
canvas open
```

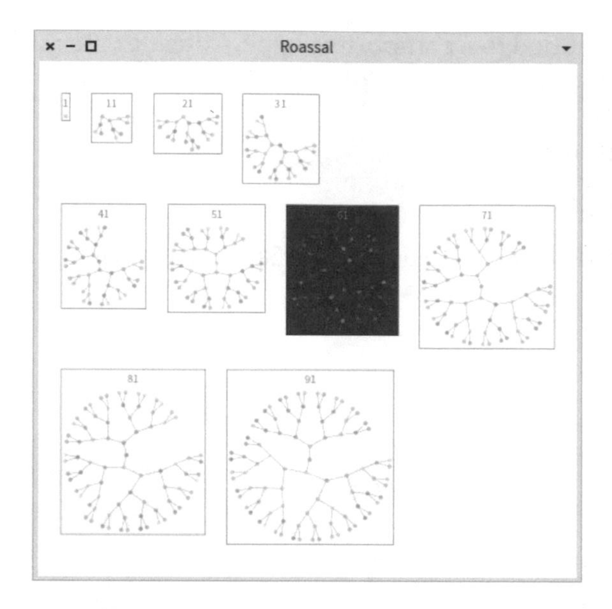

***Figure 9-4.** A composing shape can contain lines*

The group g is provided to the Line Builder using canvas:. This has the effect of making the builder generate lines in the group g.

Model Object in Composite

The RSComposite class describes a shape that other shapes can be added to. Under the hood, sending asShape creates a RSComposite object. Consider the following code:

```
c := RSCanvas new.

compoShape := RSComposite model: 'Hello' forIt: [ :composite :title |
    composite add: (RSLabel text: title).
    composite add: (RSCircle new size: 20).
    RSVerticalLineLayout new alignCenter; on: composite shapes.
    composite @ RSDraggable ].
c add: compoShape.

c open
```

Composite shapes can have a model and can be used by Line Builder. Consider the following example (see Figure 9-5):

```
classes := ByteArray withAllSubclasses.

c := RSCanvas new.
boxes := RSComposite models: classes forEach: [ :composite :cls |
    composite add: (RSLabel new text: cls name).
    composite add: (RSBox new size: cls numberOfLinesOfCode sqrt + 5).
    RSVerticalLineLayout new alignCenter; on: composite children.
    composite @ RSDraggable ].
c addAll: boxes.

lb := RSLineBuilder line.
lb withVerticalAttachPoint.
lb shapes: boxes.
lb connectFrom: #superclass.

RSTreeLayout on: c nodes.
c @ RSCanvasController.
c open
```

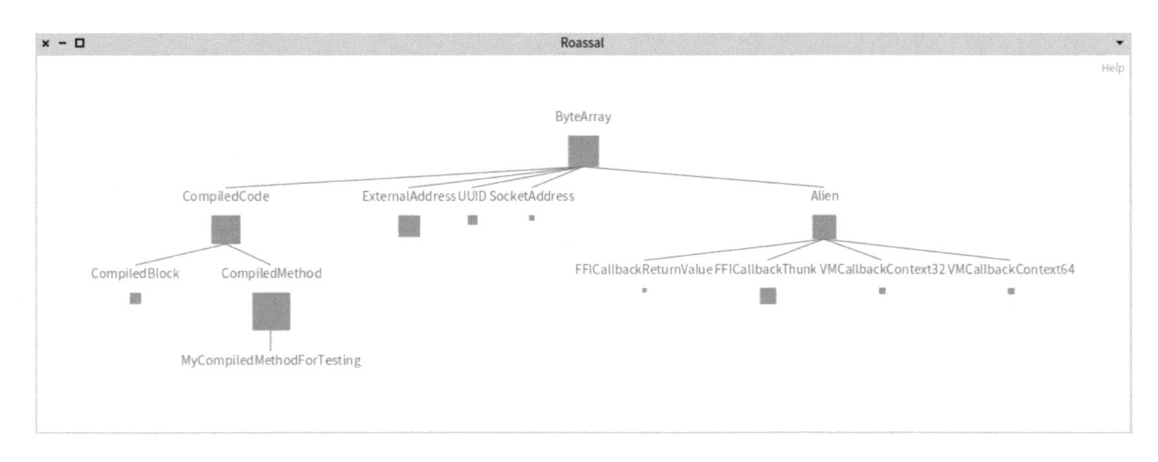

Figure 9-5. *Composing shapes*

The models:forEach: method takes two arguments:

- A collection of objects, in which each object is used as a model of a new composite shape.
- A block expecting two arguments: the new composite shape and the model object. The block is meant to fill the composite shape for a given model.

Labels Part of a Composition

In Roassal, a shape can augmented with a label in two different ways. Either by composing shapes (as you saw previously), or by using an interaction (RSLabeled). Consider the equivalent of the script given here, which uses RSLabeled instead of the composition (see Figure 9-6):

```
classes := ByteArray withAllSubclasses.

c := RSCanvas new.
boxes := RSBox models: classes forEach: [ :s :cls |
    s size: cls numberOfLinesOfCode sqrt + 5.
    s @ RSDraggable ].
c addAll: boxes.

boxes @ RSLabeled.

lb := RSLineBuilder line.
lb withVerticalAttachPoint.
lb shapes: boxes.
lb connectFrom: #superclass.

RSTreeLayout on: c nodes.
c @ RSCanvasController.
c open
```

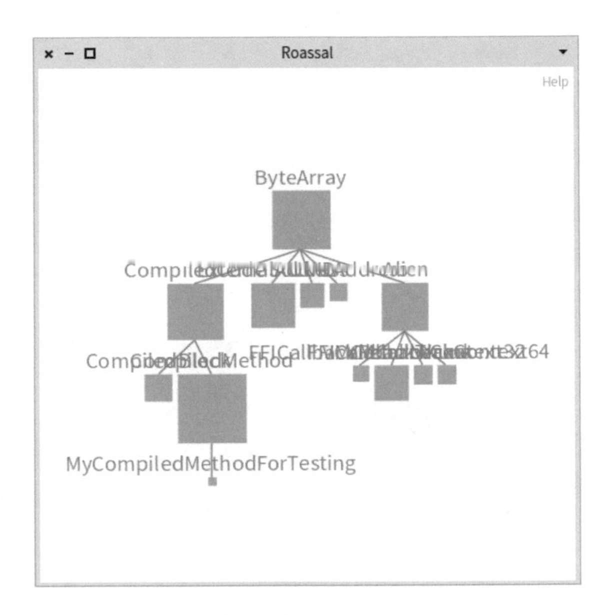

Figure 9-6. *Using RSLabeled to label shapes*

The visualization is significantly smaller. But as a consequence, all the labels are overlapping. You can use an option on RSLabeled to make them more apparent (see Figure 9-7):

```
classes := ByteArray withAllSubclasses.

c := RSCanvas new.
boxes := RSBox models: classes forEach: [ :s :cls |
    s size: cls numberOfLinesOfCode sqrt + 5.
    s @ RSDraggable ].
c addAll: boxes.

boxes @ RSLabeled highlightable.

lb := RSLineBuilder line.
lb withVerticalAttachPoint.
lb shapes: boxes.
lb connectFrom: #superclass.

RSTreeLayout on: c nodes.
c @ RSCanvasController.
c open
```

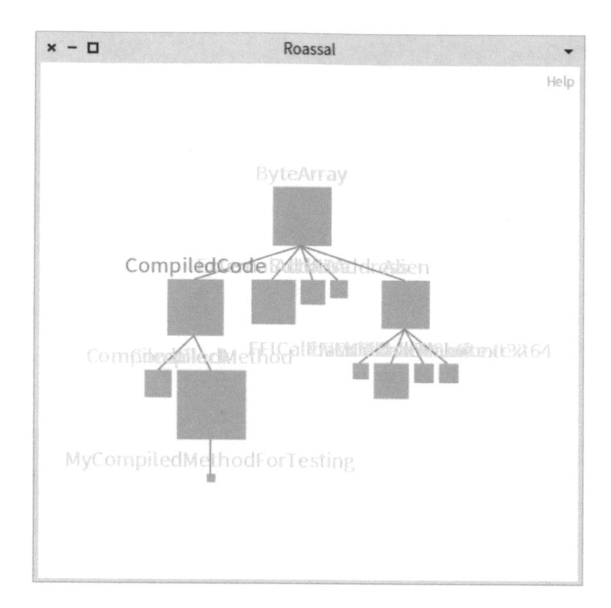

***Figure 9-7.** Highlighted RSLabeled*

A composite shape has a boundary box that encompasses all its nested shapes. As such, the layout considers this boundary box. However, when using `RSLabeled`, a label is added to the view and it does not take it into account when computing the labeled shape. Using a composite shape or an interaction depends very much on the effect that you want to have.

Labeled Circles

A frequent need when visualizing data is to have a shape with a label in its center. Consider the following script (see Figure 9-8):

```
canvas := RSCanvas new.

1 to: 30 do: [ :aNumber |
    labeledCircle := { RSCircle new size: 30 . RSLabel new color: Color
    black; text: aNumber } asGroup asShapeFor: aNumber.
    canvas add: labeledCircle ].

lb := RSLineBuilder line.
lb withBorderAttachPoint.
lb shapes: canvas nodes.
lb connectFrom: [ :v | v // 2 ].
```

```
RSTreeLayout on: canvas nodes.
canvas @ RSCanvasController.
canvas open
```

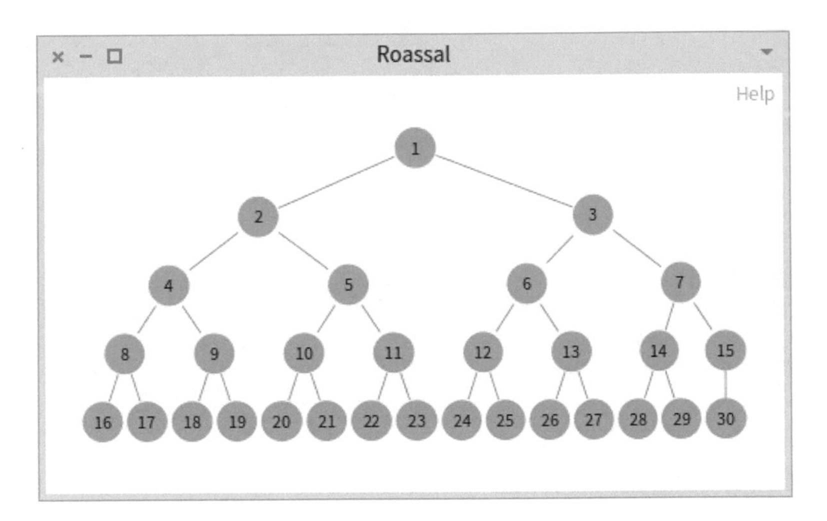

Figure 9-8. *Labeled circles*

The Pharo syntax `{ ... }` creates an array, which is converted to a `RSGroup` by sending `asGroup` to it. The `asShapeFor:` message creates a composite shape and provides a number as a model. This model is then used by Line Builder.

What Have You Learned in This Chapter?

Shape composition is an important asset of Roassal. A non-trivial visualization will likely involve composed shapes as a way to represent expressive shapes. This chapter highlights the concept of a composite shape as follows:

- Simply sending `asShape` to a group converts it to a composite shape, ready to be added to a canvas.
- A composite shape can have a border and a padding, which is useful to delimit its border.
- A composite shape may have an object model, the same way that a non-composed shape does.
- An alternative way to label shapes is using the `RSLabeled` class.

CHAPTER 10

Normalizing and Scaling Values

When building a visualization, it is common to map numerical values (e.g., the result of some metrics) or properties to visual attributes. This mapping allows you to compare properties of represented elements easily and visually. Normalizers and scales are essential components of Roassal and provide an API to map a metric or a property to a visual cue (e.g., size, color, and font size).

All the code provided in this chapter is available at `https://github.com/bergel/AgileVisualizationAPressCode/blob/main/02-06-Normalizers.txt`.

Normalizing Shape Size

Normalizing the size of shapes refers to associating a particular value or property with the size of the shape. All the normalizers are accessible from the `RSNormalizer` class. Consider the following example (see Figure 10-1):

```
values := #(3 6 10 2).

canvas := RSCanvas new.
shapes := RSEllipse models: values.
canvas addAll: shapes.

shapes @ RSLabeled.
```


A. Bergel, *Agile Visualization with Pharo*, https://doi.org/10.1007/978-1-4842-7161-2_10

```
RSNormalizer size
    shapes: shapes;
    normalize.

RSHorizontalLineLayout new alignMiddle; on: shapes.

canvas zoomToFit.
canvas open
```

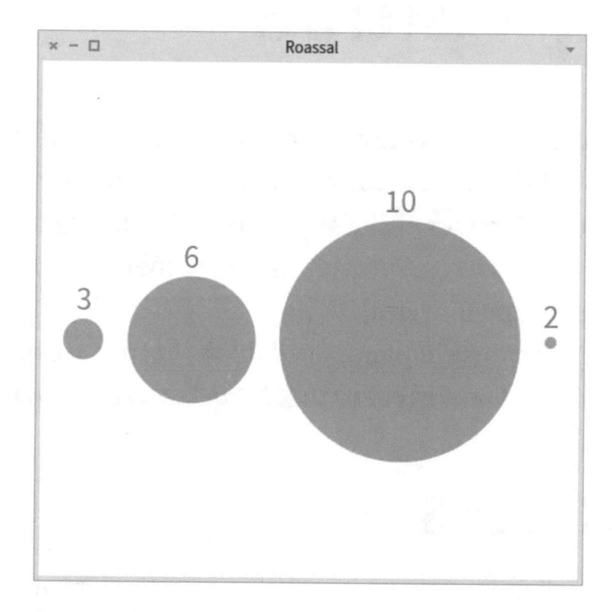

Figure 10-1. *Normalizing the shape of some circles*

The `values` variable contains four numbers. Each number is represented as a circle. You use a normalizer to make the size of the shape represent the number. The `RSNormalizer size` expression creates a normalizer to normalize the shape's size. The shapes are provided to the normalizer using `shapes:` and the normalization is carried out by sending `normalize`.

Note that the `RSNormalizer size shapes: shapes; normalize` expression makes sense *only* because the model of each shape is a number. The `RSEllipse models: values` expression creates circles with a number as a model since the `values` variable is a collection of numbers.

The range of the normalization can be manually set using `from:` and `to:`. Consider this revision (see Figure 10-2):

```
values := #(3 6 10 2).

canvas := RSCanvas new.
shapes := RSEllipse models: values.
canvas addAll: shapes.

shapes @ RSLabeled.

RSNormalizer size
    shapes: shapes;
    from: 10;
    to: 30;
    normalize.

RSHorizontalLineLayout new alignMiddle; on: shapes.

canvas zoomToFit.
canvas open
```

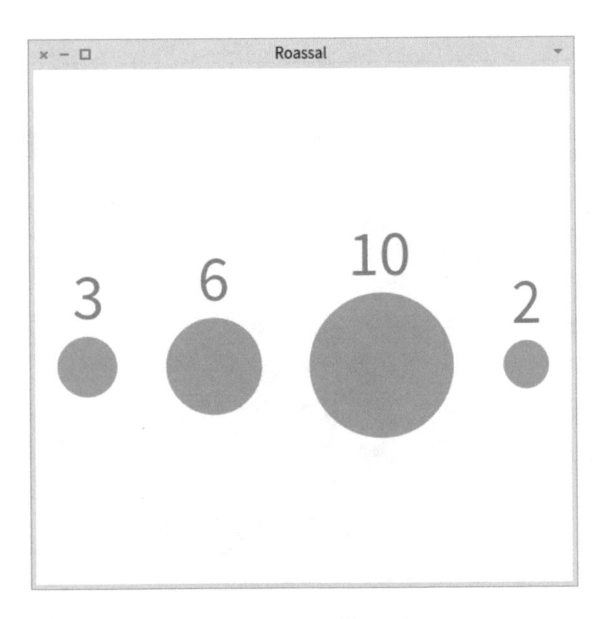

Figure 10-2. *Setting the range of the normalization*

Values in the values variable are mapped to sizes ranging from 10 to 30 pixels. The smallest number (2) contained in the values variable is 10 pixels, while the greatest value (10) is 30 pixels.

When applying a normalization, a transformation could be performed. A transformation may have many relevant properties, including coping with disparate values and diminishing outliers. Consider the following code (see Figure 10-3):

```
values := #(3 6 10 2).

canvas := RSCanvas new.
shapes := RSEllipse models: values.
canvas addAll: shapes.

shapes @ RSLabeled.

RSNormalizer size
    shapes: shapes;
    normalize: [ :aNumber | aNumber * aNumber * aNumber ].

RSHorizontalLineLayout new alignMiddle; on: shapes.

canvas zoomToFit.
canvas open
```

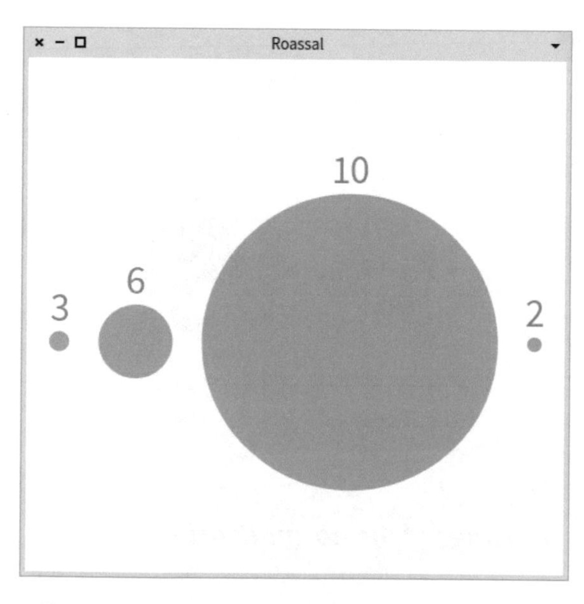

Figure 10-3. *Applying a simple transformation*

Values are raised to the power 3, which has the effect of highlighting any differences: large values appear much larger than smaller values. Conversely, differences between values may be reduced with a square root transformation (`[ :aNumber | aNumber sqrt]`) or a logarithmic transformation (`[ :aNumber | aNumber ln ]`).

The following sections describe the different normalizers offered by Roassal.

The RSNormalizer Class

The `RSNormalizer` class is the entry point to using normalizers. A normalizer is created by sending a message to the class indicating the visual attribute to normalize. Available normalizers can be obtained as follows:

- `RSNormalizer color` to normalize the color of the shape
- `RSNormalizer fontSize` to normalize the font size (note that the specified shapes must be labels)
- `RSNormalizer height` to normalize the height
- `RSNormalizer position` to normalize the position of the shapes
- `RSNormalizer size` to normalize the size of shape, as illustrated previously
- `RSNormalizer width` to normalize the width

Shapes to be normalized have to be provided using `shapes:`. The range of the normalization can be specified using `to:` and `from:`. These two methods accept a value that depends on the accepted values for the attribute to normalize; it accepts numbers, points, and color, depending on the usage.

In the previous example, the normalization is performed using `normalize`. This is a particular case in which the shapes to be normalized have a number as a model. You can use `normalize:` instead of providing a block that transforms a model object into the value to normalize, as illustrated next.

You can try the different normalizers by simply replacing `RSNormalizer size` in the previous with by one of the normalizers given here.

Combining Normalization

Several normalizers can be applied to the same shapes. Consider the following example, which gives a visual representation of five translations of "hello" (see Figure 10-4):

```
translations := #('Bonjour' 'Guten Morgen' 'Hola' 'Buongiorno' 'Dia dhuit').

canvas := RSCanvas new.
shapes := RSBox models: translations.
canvas addAll: shapes.

shapes @ RSLabeled.

RSNormalizer height
    shapes: shapes;
    normalize: [ :text | (text select: #isVowel) size ].

RSNormalizer width
    shapes: shapes;
    normalize: [ :text | (text reject: #isVowel) size ].

RSNormalizer color
    shapes: shapes;
     from: Color green;
     to: Color blue;
    normalize: [ :text | text size ].

RSHorizontalLineLayout new gapSize: 10; alignMiddle; on: shapes.

canvas zoomToFit.
canvas open
```

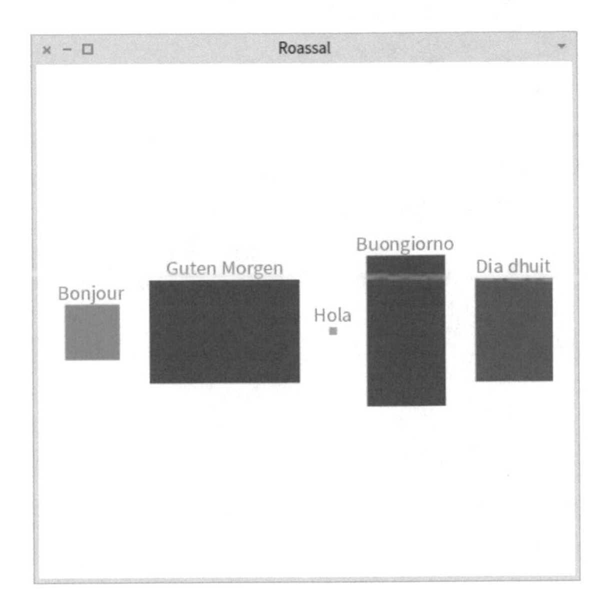

Figure 10-4. *Multiple normalizations*

Each translation is represented as a box. The height, width, and color of each box is particular to the represented word. The height indicates the number of vowel characters, the width is the number of consonant characters, and the color indicates the number of characters in the translation.

Normalizing Shape Position

Points describing the position of some shapes can be normalized. Consider the following example (see Figure 10-5):

```
classes := Collection withAllSubclasses.

canvas := RSCanvas new.

dots := RSCircle models: classes forEach: [ :s :o | s borderColor: #black ].
canvas addAll: dots.
dots @ RSPopup.

RSNormalizer size
    shapes: dots;
    from: 10;
    to: 25;
    normalize: #numberOfLinesOfCode.
```

```
RSNormalizer color
    shapes: dots;
    from: Color yellow;
    to: Color brown;
    normalize: #numberOfLinesOfCode.

canvas add: (RSArrowedLine new from: 0 @ 0; to: 0 @ -500).

RSNormalizer position
    shapes: dots;
    from: 0 @ 0;
    to: 500 @ -500;
    normalize: [ :aClass | aClass numberOfMethods @ aClass
    numberOfLinesOfCode ].

canvas add: (RSArrowedLine new from: 0 @ 0; to: 0 @ -500).
canvas add: (RSArrowedLine new from: 0 @ 0; to: 500 @ 0).
canvas zoomToFit.

canvas open
```

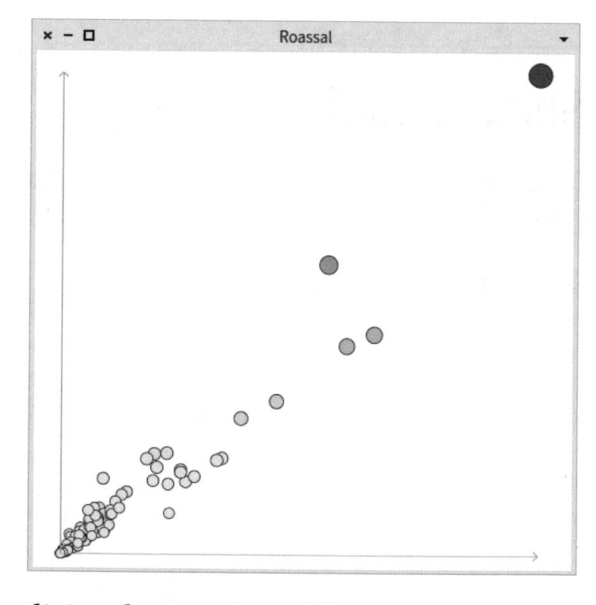

Figure 10-5. *Normalizing the position of shapes*

The script simulates a scatterplot by normalizing some shapes based on some metrics. Each circle is a class. The number of methods is mapped to the x-axis and the number of lines of code is mapped to the y-axis, the size of the circle, and the color. A class with many lines of code appears as a large brown circle. A class with relatively few lines of code is represented as a small yellow circle. The bloc provided to the `RSNormalizer position` normalizer returns a position and it is used to locate each class in the scatterplot.

Line Width

The width of the line can also be normalized by simply using `RSNormalizer width` and providing lines to the normalizer. Consider the following example (see Figure 10-6):

```
canvas := RSCanvas new.
shapes := RSEllipse models: Collection withAllSubclasses.
shapes @ RSDraggable @ RSPopup.

canvas addAll: shapes.
RSNormalizer color
    from: (Color gray alpha: 0.5);
    to: (Color red alpha: 0.5);
    shapes: shapes;
    normalize: #numberOfMethods.
RSNormalizer size
    shapes: shapes;
    normalize: #numberOfMethods.
RSLineBuilder line
    canvas: canvas;
    withBorderAttachPoint;
    connectFrom: #superclass.

RSNormalizer width
    shapes: canvas lines;
    from: 2;
    to: 10;
```

```
    normalize: [ :association |
        (association key selectors intersection: association value
        selectors) size ].

RSTreeLayout on: shapes.

canvas @ RSCanvasController.
canvas open
```

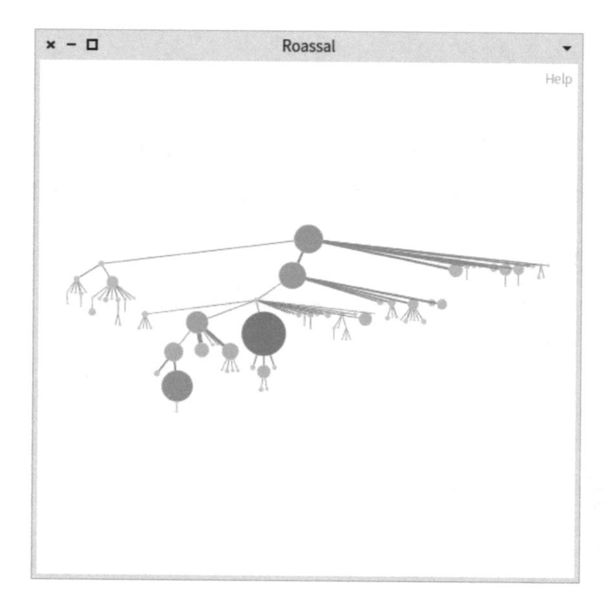

Figure 10-6. *Normalizing line width*

The example visualizes the collection class hierarchy. Each class is represented as a circle. The number of methods defined in the class is visually represented by both the size of the circle and its color, which ranges from translucent gray to translucent red. The translucence is set using the `alpha:` method, which expects a float number ranging from 0 to 1 to indicate the degree of opacity (values lower than 1 make the shape translucent).

Lines are defined using a line builder, and each line connects a superclass of a class to that class. The lines can be accessed using the `canvas lines` expression, which simply returns a group of lines.

The line width is normalized using `RSNormalizer width` on the lines. The normalization is performed using a one-argument bloc. The model of a line is an association having:

- The model of the starting shape as the key
- The model of the ending shape as the value

The line width indicates the number of similar method names between the superclass and its subclass. A thick line indicates that many methods of the superclass are directly redefined (i.e., overridden) in the subclass. This examples illustrates how to map an arbitrary metric to line thickness.

Scaling

A normalizer uses a scale object to map values from one scale to another. Many scales are available and the `NSScale` class is the root of a class hierarchy. As this section illustrates, it can be beneficial to directly employ scales.

Consider the following example:

```
s := NSScale linear.
s range: { 5 . 10 }.
s scale: 0. "=> 5"
s scale: 0.5. "=> 7.5"
s scale: 1. "=> 10"
```

The `scale:` method takes an object as an argument—a number in this example. The value provided to `range:` is an array of two or more values to which the domain is mapped. The domain used in this example is the default one, which is `#(0 1)`. This can be verified by evaluating `NSScale linear domain`, which returns `#(0 1)`.

Consider the following scale that simply maps the domain value `[-10, -5]` to the values ranging from 10 to 5:

```
s := NSScale linear.
s domain: { -10 . -5 }.
s range: { 10 . 5 }.
s scale: -5. "=> 5"
s scale: -7.5. "=> 7.5"
s scale: -10. "=> 10"
```

Numerical values can also be mapped to colors. Consider this example:

```
s := NSScale linear.
s range: { Color white . Color black }.
s scale: 0.5  "=> Color gray"
```

A larger example illustrates color mapping (see Figure 10-7):

```
values := 0 to: 1 by: 0.1.
scale := NSScale linear range: { Color blue . Color red }.
canvas := RSCanvas new.
shapes := RSCircle models: values forEach: [ :shape :number |
                    shape color: (scale scale: number) ].
shapes @ RSPopup.
canvas addAll: shapes.
RSFlowLayout on: shapes.
canvas zoomToFit.
canvas open
```

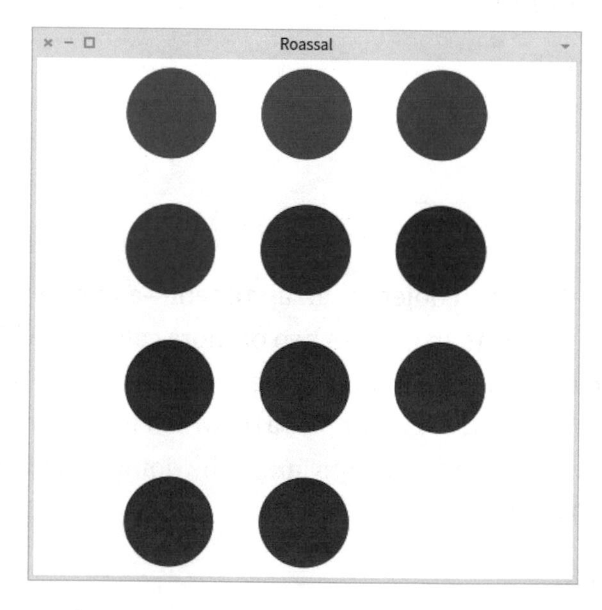

Figure 10-7. *Scaling colors*

The color is obtained with the `scale scale: number` expression, which returns a color ranging from blue to red. As you can see, the example manually simulates a normalizer. One advantage that scales have over normalizers is when considering multiple values for the domain and range. For example, consider a variation of the previous example (see Figure 10-8):

```
values := 0 to: 1 count: 20.
scale := NSScale linear
               range: { Color blue . Color gray . Color red };
               domain: { 0 . 0.5 . 1 }.
canvas := RSCanvas new.
shapes := RSCircle models: values forEach: [ :shape :number |
                    shape color: (scale scale: number) ].
shapes @ RSPopup.
canvas addAll: shapes.
RSFlowLayout on: shapes.
canvas zoomToFit.
canvas open
```

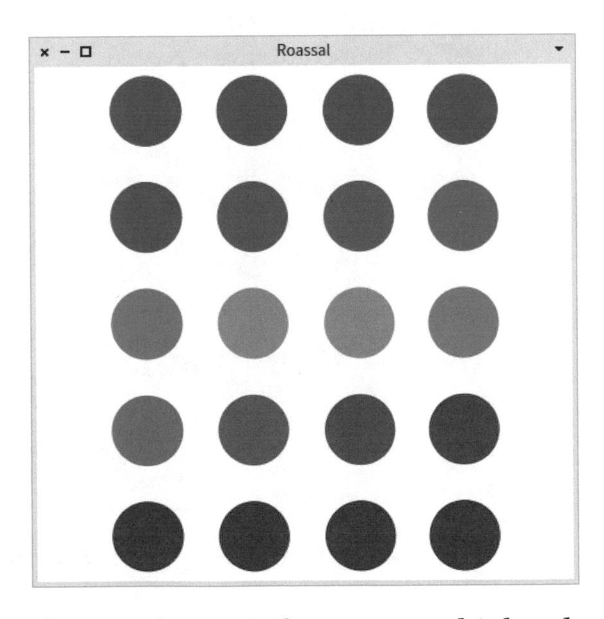

Figure 10-8. *The scaling can be carried out over multiple values*

The example maps numerical values to a collection of three colors—blue, gray, and red. Since the domain is specified as `{ 0 . 0.5 . 1 }`, blue represents `0`, gray represents `0.5`, and red represents `1`. Note that the value of gray does not have to be `0.5` (the middle between 0 and 1, the two extremities of the domain).

Multiple scales can be used on the same shapes (see Figure 10-9):

```
sc := NSScale linear
 range: { #white . #gray . #red };
 domain: { 0 . 1 . 2}.

ss := NSScale linear
 range: { 80 . 30 . 80 };
 domain: { 0 . 1 . 2}.

c := RSCanvas new.
0 to: 2 by: 0.1 do: [ :i |
 s := RSBox new width: 30; height: (ss scale: i); model: i; color: (sc
 scale: i).
 c add: s ].

RSHorizontalLineLayout new alignMiddle; on: c shapes.
c zoomToFit.
c open
```

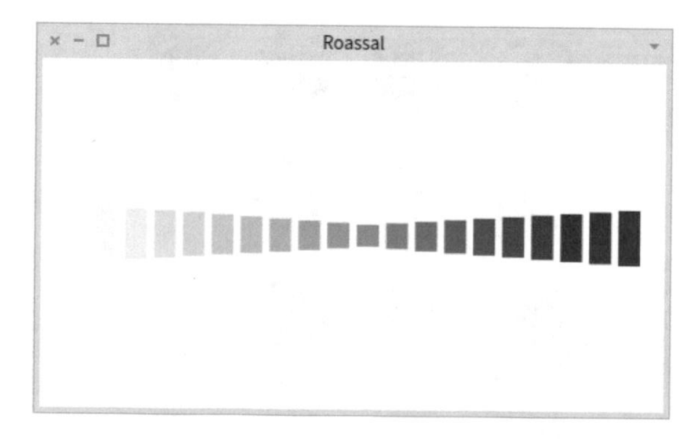

Figure 10-9. *Colors and height are mapped through multiple range values*

Roassal supports ordinal scaling, which is useful for assigning a color to a particular object (instead of a numerical value, as you have just seen). Consider the following example (see Figure 10-10):

```
c := RSCanvas new.
scale := NSScale ordinalColored: 3.

r := Random seed: 42.
9 timesRepeat: [
    txt := { 'hello' . 'bonjour' . 'Guten Morgen' } atRandom: r.
    g := RSGroup new.
    g add: (RSBox new width: 100; height: 20; model: txt; color: (scale
    scale: txt)).
    g add: (RSLabel new text: txt).
    c add: g asShape ].

RSGridLayout new lineItemsCount: 3; on: c shapes.
c @ RSCanvasController.
c open
```

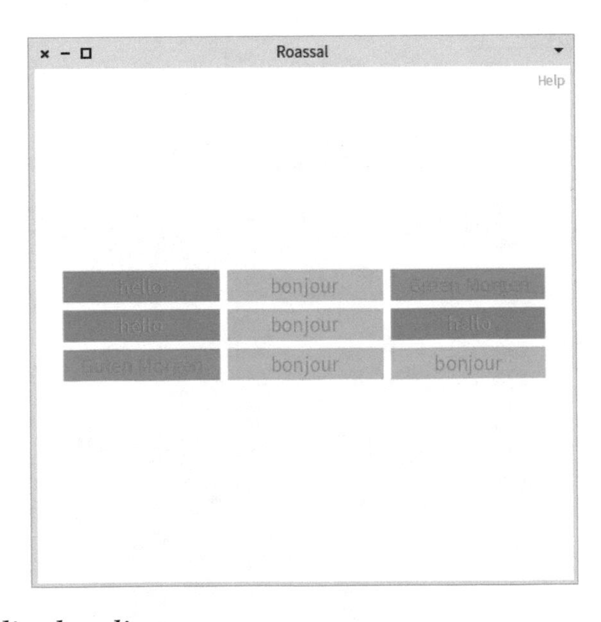

Figure 10-10. *Ordinal scaling*

The `NSScale ordinalColored: 3` expression creates an ordinal scale with three different colors. The `colors scale: txt` expression returns one of the three colors assigned to the object `txt`. A color can be used for more than one object when there are more objects provided to `scale:` than the number of colors provided to `ordinalColored:`. The `ordinalColored:` method creates a random palette of a particular size.

Several color palettes are provided on the class side of `NSScale`. Consider the following example (see Figure 10-11):

```
canvas := RSCanvas new.
scale := NSScale category20c.

r := Random seed: 42.
81 timesRepeat: [
    aNumber := r nextInt: 10.
    circle := RSCircle new size: 30; color: (scale scale: aNumber).
     label := RSLabel new color: Color black; text: aNumber.
    composite := { circle . label } asGroup asShapeFor: aNumber.
    canvas add: composite ].

RSGridLayout new lineItemsCount: 9; on: canvas nodes.

canvas @ RSCanvasController.
canvas open
```

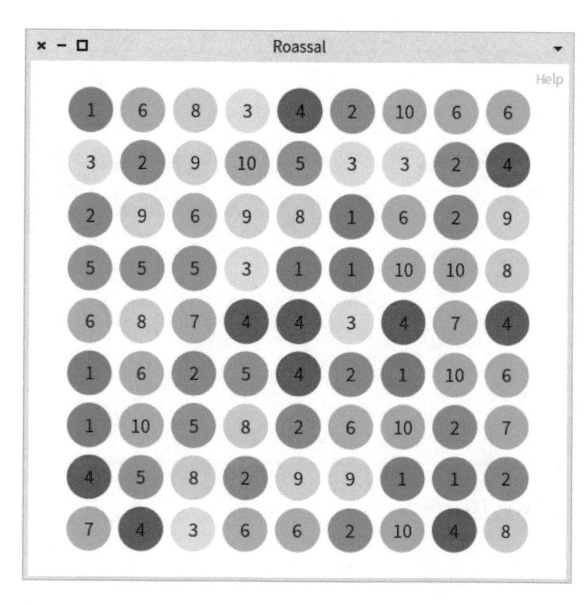

Figure 10-11. *Using the category20c palette*

The script associates a number to a particular color. Two circles representing the same number have the same color. In addition to using the predefined palettes of Roassal, you can manually build a color palette and provide it directly using `range:`. Consider this example (see Figure 10-12):

```
canvas := RSCanvas new.

palette := { 'e66e33' . 'fca300' . 'efdb2b' . 'a3bd21' . '009991' }
            collect: [ :hexString | Color fromHexString: hexString ].

scale := NSScale ordinal range: palette.

r := Random seed: 42.
100 timesRepeat: [
    aNumber := r nextInt: 5.
    circle := RSCircle new model: aNumber.
    circle @ RSPopup.
    circle color: (scale scale: aNumber).
    canvas add: circle ].

RSFlowLayout on: canvas nodes.

canvas @ RSCanvasController.
canvas open
```

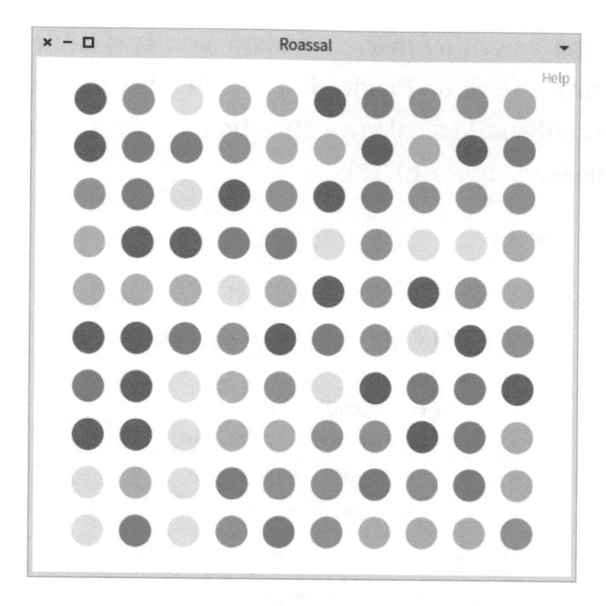

Figure 10-12. *Using a manually defined palette*

A hexadecimal color description is converted to a color using the `Color fromHexString: hexString` expression. The color palette used in the previous example is called "Summer Stripes Color Palette" and it was obtained from `http://www.color-hex.com`.

What Have You Learned in This Chapter?

This chapter covered the use of normalizers and scales, which are a necessary component to map numerical values to visual attributes. The chapters covered:

- The normalization mechanism of visual attributes, in particular the size, line width, color, and position of shapes
- Scaling over multiple values

CHAPTER 11

Interactions

An interactive visualization allows additional information to be presented when a user triggers actions through the mouse or keyboard. Roassal proposes a number of pluggable and extensible interactions, presented in this chapter.

All the code provided in this chapter is available at https://github.com/bergel/AgileVisualizationAPressCode/blob/main/02-07-Interactions.txt.

Useful Interactions

So far, you have used three main interactions: RSPopup, RSDraggable, and RSCanvasController. Consider the following example, which involves these three interactions:

```
canvas := RSCanvas new.

shapes := RSCircle models: (1 to: 100).
canvas addAll: shapes.
shapes @ RSPopup @ RSDraggable.
RSGridLayout on: shapes.

canvas @ RSCanvasController.
canvas open
```

An interaction can be set by using the @ message and the interaction as an argument. RSPopup enables shapes with models to have a popup window that shows the model as a string representation. The window appears by moving the mouse above it. In this example, each number is displayed in the popup window since each circle

A. Bergel, *Agile Visualization with Pharo*, https://doi.org/10.1007/978-1-4842-7161-2_11

has a number as a model. RSDraggable makes the shape draggable with the mouse. RSCanvasInteraction is an interaction for the canvas that augments the canvas with the following interactions:

- Zooming in/out using the mouse wheel or the I and O keys.
- Moving the canvas by dragging it with the mouse or using the arrow keys.
- Searching for particular shapes using the S and R keys.
- Maximizing the window using the M key.
- Adding a Help menu that summarizes the available shortcuts.

These three interactions are frequently used in this book.

Using Any Shape in a Popup

The default behavior of the RSPopup class is to create popup windows with labels in them. However, the popup may be freely customized. For example (see Figure 11-1):

```
canvas := RSCanvas new.

shapes := RSLabel models: (1 to: 10).
canvas addAll: shapes.
RSGridLayout on: shapes.

popup := RSPopup new.
popup shapeBuilder: [ :aNumber |
    | g |
    g := RSGroup new.
    g addAll: (RSCircle models: (1 to: aNumber)).
    RSGridLayout on: g.
    g asShape
        color: Color white darker darker;
        padding: 5.
].
```

```
shapes @ popup.
shapes @ RSHighlightable red.

canvas zoomToFit.
canvas open
```

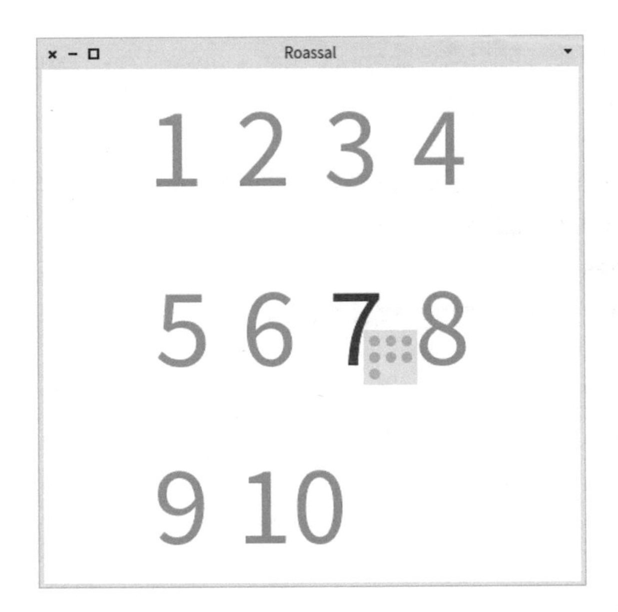

Figure 11-1. *Customized popup*

The shapeBuilder: method lets you customize the way the popup is generated. The method accepts as an argument a block that takes the model object as an argument. It has to return a shape or a composite shape. Consider a slightly more sophisticated example in the following code (see Figure 11-2):

```
packages := RPackageOrganizer default packages select: [ :rPak | rPak
packageName beginsWith: 'Roassal3' ].

canvas := RSCanvas new.
shapes := RSBox models: packages.
canvas addAll: shapes.

shapes @ RSDraggable.
```

```
RSNormalizer size
    shapes: shapes;
    from: 5; to: 20;
    normalize: [ :rPackage | rPackage definedClasses size ].

popup := RSPopup new.
popup shapeBuilder: [ :package |
    | g nodes lbl |
    g := RSGroup new.
    nodes := RSBox models: package definedClasses.
    nodes color: Color blue.
    g addAll: nodes.
    tlb := RSLineBuilder orthoVertical.
    tlb canvas: g.
    tlb shapes: nodes.
    tlb connectFrom: #superclass.
    RSTreeLayout on: nodes.
    lbl := RSLabel text: package packageName.
    g add: lbl.
    RSLocation move: lbl above: nodes.
    g asShape
        color: Color white darker translucent;
        borderColor: #black;
        padding: 3.
].
shapes @ popup.

RSFlowLayout on: shapes.
canvas @ RSCanvasController.
canvas open
```

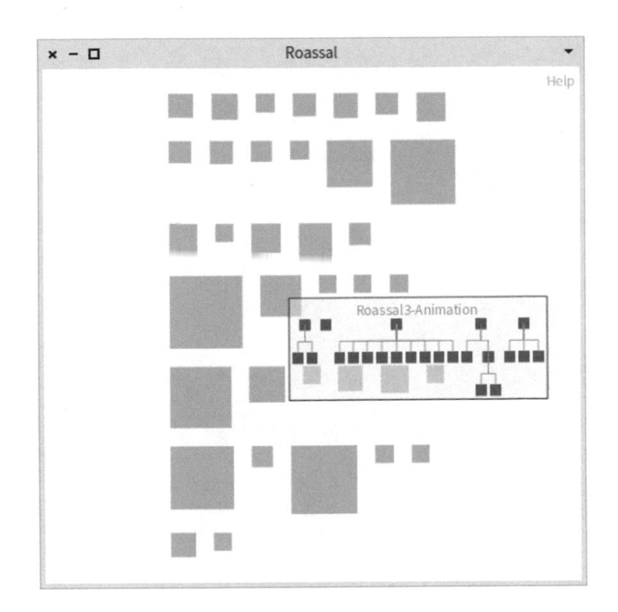

Figure 11-2. *Customized popup*

The visualization shows the Pharo packages that define the Roassal3 application. The size of each package indicates the number of classes it contains. Moving the mouse above a package shows the classes and their inheritance in a popup window. The block provided to the shape builder adds nodes, lines, and a label giving the name of the package.

RSLabeled

The `RSLabeled` interaction is useful to set a title to some shapes. Consider the following code (see Figure 11-3):

```
canvas := RSCanvas new.

shapes := RSCircle models: (1 to: 9).
canvas addAll: shapes.
shapes @ RSLabeled.
RSGridLayout new gapSize: 40; on: shapes.

canvas @ RSCanvasController.
canvas open
```

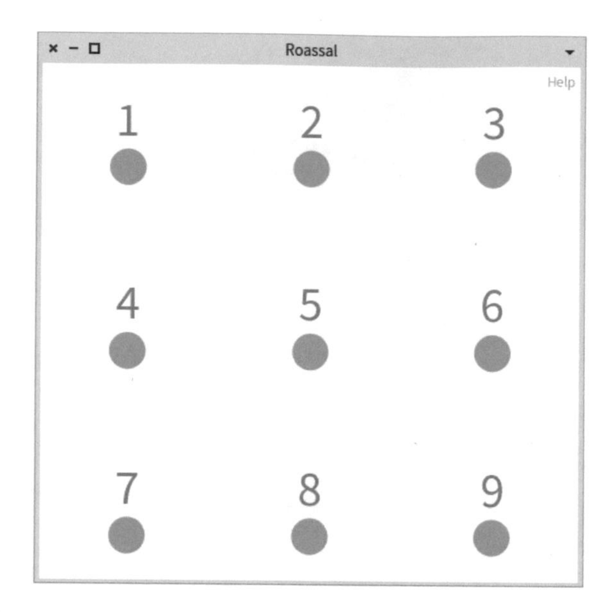

Figure 11-3. *Adding an RSLabeled interaction*

The RSLabeled interaction associates a label to a shape. By default, the label is located above the shape, as shown in Figure 11-3. Note that the interaction does not create a composite shape, and as a consequence, the layout does not take the label into account (as explained in the chapter about shape composition).

The label produced by RSLabeled can be customized using text (see Figure 11-4):

```
canvas := RSCanvas new.

shapes := RSCircle models: (1 to: 9).
canvas addAll: shapes.
shapes @ (RSLabeled new text: [ :aModel | 'value = ', aModel asString ]).
RSGridLayout new gapSize: 20; on: shapes.

canvas @ RSCanvasController.
canvas open
```

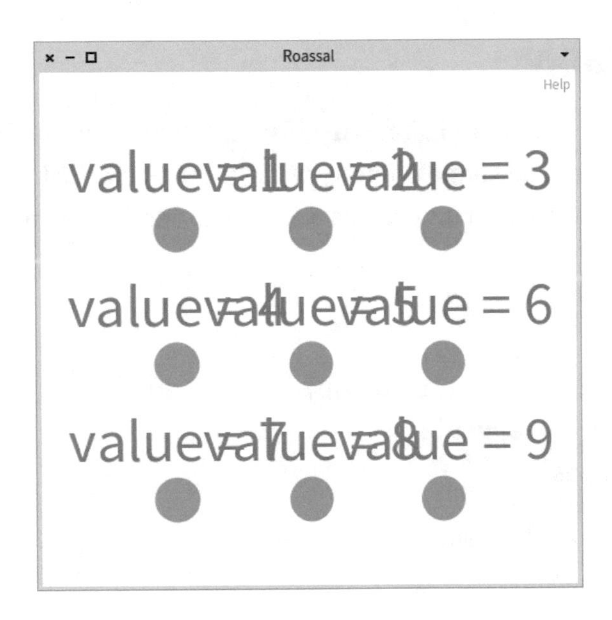

Figure 11-4. *Overlapping labels*

You can make `RSLabeled` highlightable to ease reading. Consider this revision of the previous script:

```
canvas := RSCanvas new.

shapes := RSCircle models: (1 to: 9).
canvas addAll: shapes.
shapes @ (RSLabeled new text: [ :aModel | 'value = ', aModel asString ];
highlightable).
RSGridLayout new gapSize: 20; on: shapes.

canvas @ RSCanvasController.
canvas open
```

By setting `RSLabeled` as `highlightable`, the color of the label, initially very light gray, is darkened when the mouse is above the shape. The next section deals with highlightable shapes.

RSHighlightable

The `RSHighlightable` interaction allows users to highlight shapes when the mouse cursor is above a shape. The shape pointed by the mouse may be highlighted, but other shapes that are semantically related can as well. Consider this simple example (see Figure 11-5):

```
canvas := RSCanvas new.

shapes := RSEllipse models: (1 to: 30).
shapes @ RSPopup @ RSDraggable.
canvas addAll: shapes.

shapes @ RSHighlightable red.
RSGridLayout on: shapes.

canvas @ RSCanvasController.

canvas open
```

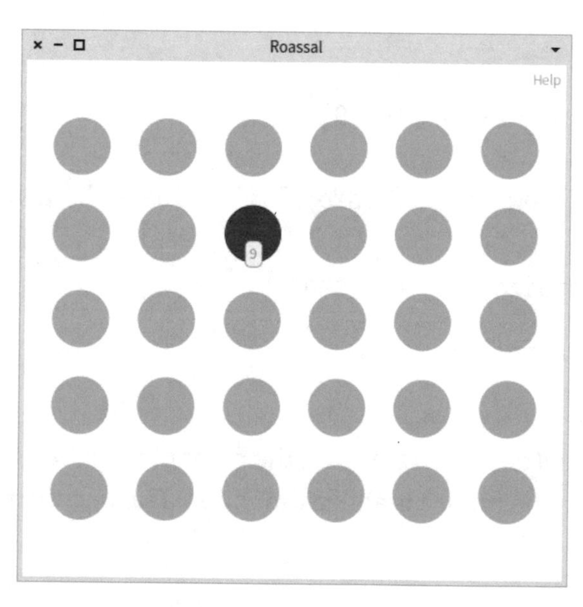

Figure 11-5. *The effect of highlighting a shape*

The interaction is set using the `shapes @ RSHighlightable red` expression. The color used for highlighting can be manually set using `highlightColor:`. For example:

```
canvas := RSCanvas new.

shapes := RSEllipse models: (1 to: 30).
shapes @ RSPopup @ RSDraggable.
canvas addAll: shapes.

shapes @ (RSHighlightable new highlightColor: Color green).
RSGridLayout on: shapes.

canvas @ RSCanvasController.

canvas open
```

Shapes to be highlighted can be specified using `highlightShapes:`. Consider the following example (see Figure 11-6):

```
canvas := RSCanvas new.

shapes := RSEllipse models: (1 to: 30).
shapes @ RSPopup @ RSDraggable.
canvas addAll: shapes.

highlightable := RSHighlightable new.
highlightable highlightColor: Color green.
highlightable highlightShapes: [ :aShape |
    shapes select: [ :s | (-1 to: 1) includes: (s model - aShape model)  ] ].
shapes @ highlightable.
RSGridLayout on: shapes.

canvas @ RSCanvasController.

canvas open
```

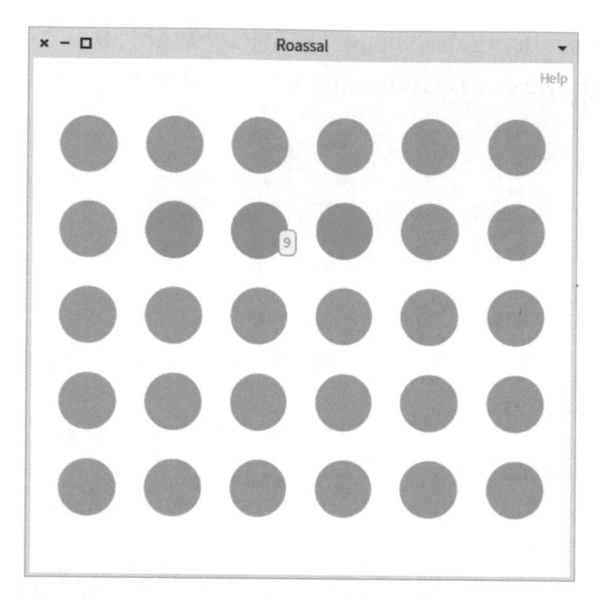

Figure 11-6. *Selecting shapes to be highlighted*

Lines can be highlighted using the `withEdges` method. For example (see Figure 11-7):

```
numberOfNodes := 25.
numberOfLines := 90.
r := Random seed: 42.
graph := Dictionary new.

1 to: numberOfNodes do: [ :aNode |
    graph at: aNode put: Set new ].

numberOfLines timesRepeat: [
    fromNode := r nextInteger: numberOfNodes.
    toNode := r nextInteger: numberOfNodes.
    (graph at: fromNode) add: toNode ].

canvas := RSCanvas new.
nodes := RSCircle models: (1 to: numberOfNodes).
nodes @ RSDraggable @ RSPopup.
canvas addAll: nodes.

highlightable := RSHighlightable new.
highlightable highlightColor: Color red.
```

```
highlightable withEdges.
nodes @ highlightable.

lb := RSLineBuilder line.
lb canvas: canvas.
lb withBorderAttachPoint.
lb makeBidirectional.
lb moveBehind.
lb objects: (1 to: numberOfNodes).
lb connectToAll: [ :aNumber | graph at: aNumber ].

RSForceBasedLayout new charge: -300; on: nodes.
canvas @ RSCanvasController.
canvas open
```

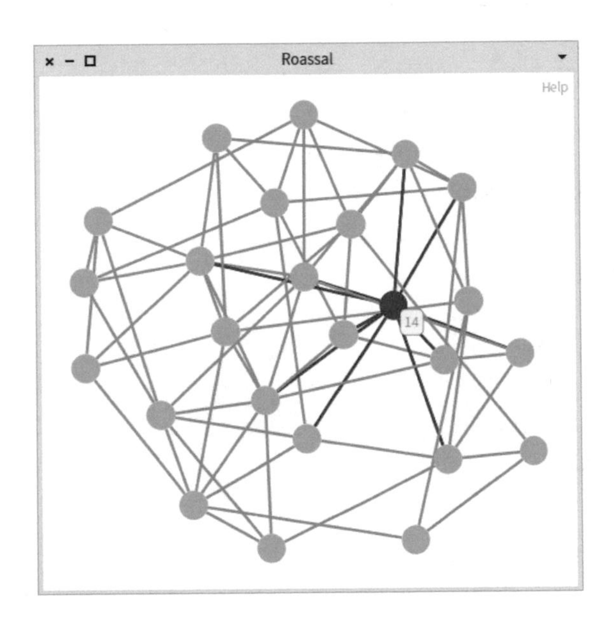

Figure 11-7. *Highlighting edges*

Several highlightable interactions can be combined. For example (see Figure 11-8):

```
numberOfNodes := 25.
numberOfLines := 90.
r := Random seed: 42.
graph := Dictionary new.
```

```
1 to: numberOfNodes do: [ :aNode |
    graph at: aNode put: Set new ].

numberOfLines timesRepeat: [
    fromNode := r nextInteger: numberOfNodes.
    toNode := r nextInteger: numberOfNodes.
    (graph at: fromNode) add: toNode ].

canvas := RSCanvas new.
nodes := RSCircle models: (1 to: numberOfNodes).
nodes @ RSDraggable @ RSPopup.
canvas addAll: nodes.

highlightable := RSHighlightable new.
highlightable highlightColor: Color red.
highlightable withEdges.
nodes @ highlightable.

highlightable2 := RSHighlightable new.
highlightable2 highlightColor: Color blue.
highlightable2 withConnectedShapes.
nodes @ highlightable2.

lb := RSLineBuilder line.
lb canvas: canvas.
lb withBorderAttachPoint.
lb makeBidirectional.
lb moveBehind.
lb objects: (1 to: numberOfNodes).
lb connectToAll: [ :aNumber | graph at: aNumber ].

RSForceBasedLayout new charge: -300; on: nodes.
canvas @ RSCanvasController.
canvas open
```

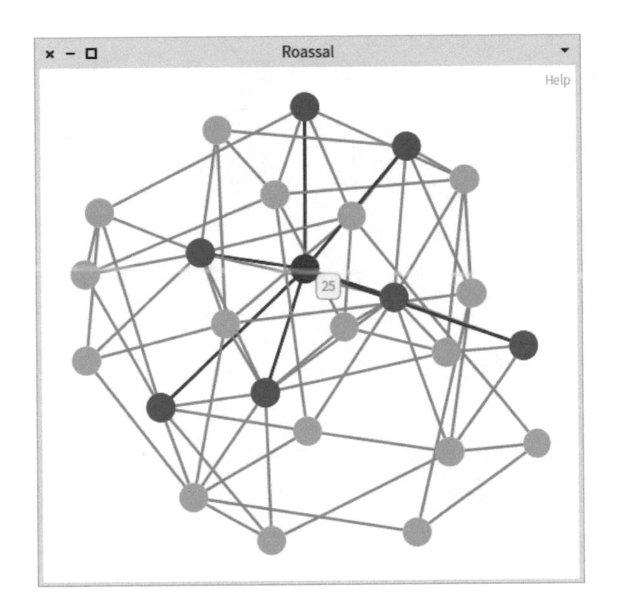

Figure 11-8. *Combining highlightable interactions*

Two highlightable interactions are set on the shapes contained in `nodes`. The first interaction, `highlightable`, highlights the edges in red, while the second interaction, `highlightable2`, highlights the connected shapes.

What Have You Learned in This Chapter?

This chapter covered a number of essential interactions provided by Roassal. These interactions are intended to let you collect more relevant information through simple actions triggered by moving the mouse. The chapter covers the following:

- `RSPopup` to obtain a popup window
- `RSDraggable` to make shapes draggable
- `RSLabeled` to label shapes
- `RSHighlightable` to highlight shapes

CHAPTER 12

Layouts

Roassal offers a number of layouts to locate shapes in the two-dimensional space provided by the canvas. Layouts have been extensively used in the previous chapters without describing how they operate and how they are configured. This chapter lists the different layouts supported by Roassal and details the layout framework.

This chapter covers the following layouts: circle, grid, flow, rectangle pack, and line. It lists layouts that are edge-driven, i.e., lines connecting to shapes as a way to structure the layout. Edge-driven layouts are tree and forces. The bridge with Graphviz is also presented.

Circle Layout

Consider the following example (see Figure 12-1):

```
nodes := (1 to: 5).

canvas := RSCanvas new.
shapes := RSCircle models: nodes.
shapes size: 30.
canvas addAll: shapes.

RSCircleLayout on: shapes.
canvas @ RSCanvasController.
canvas open
```


A. Bergel, *Agile Visualization with Pharo*, https://doi.org/10.1007/978-1-4842-7161-2_12

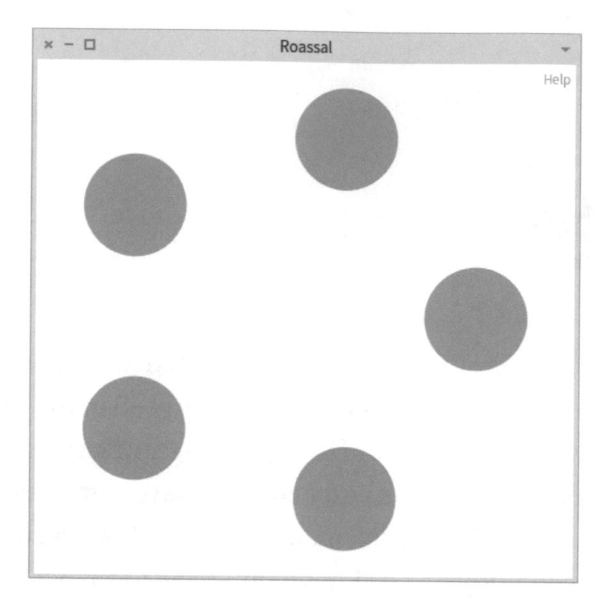

Figure 12-1. *Circle layout*

The example locates five shapes as a circle. The circle layout can be configured in many different ways. In particular:

- The initial angle used by the layout can be set using `initialAngleInDegree:`.
- The radius can be set using `radius:`, a method that takes a number of pixels as the circle radius.

The `RSCircleLayout` circle layout is adequate when all the shapes have roughly the same size. Overlap is likely to happen when there are significant disparities in the shapes' sizes. Consider the following script (see Figure 12-2):

```
nodes := (1 to: 20) asArray shuffleBy: (Random seed: 42).

canvas := RSCanvas new.
shapes := RSCircle models: nodes.
canvas addAll: shapes.
RSNormalizer size
    shapes: shapes;
    normalize.
```

```
RSCircleLayout on: shapes.
canvas @ RSCanvasController.
canvas open
```

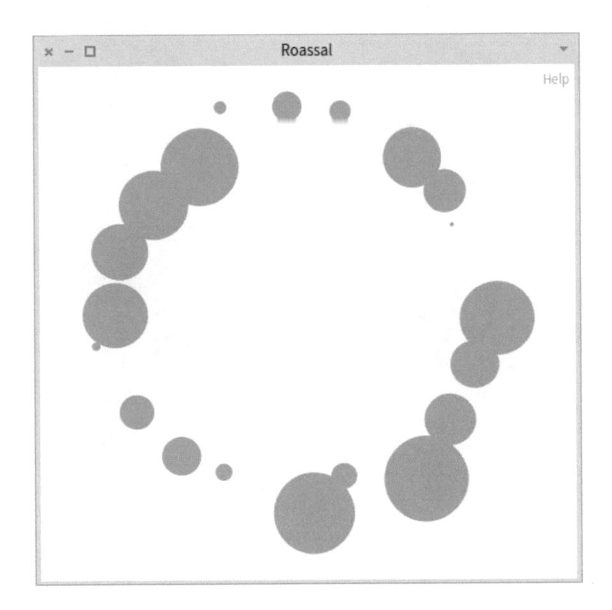

***Figure 12-2.** Overlapping shapes using a circle layout*

The circle layout locates shapes based on their center along a circle. The angle between shapes from the circle center is constant, and because of this, overlap may happen. To avoid overlapping shapes, a variant of RSCircleLayout is proposed, called RSEquidistantCircleLayout.

The RSEquidistantCircleLayout class ensures there is a constant space between circles. Consider the following code (see Figure 12-3):

```
nodes := (1 to: 20) asArray shuffleBy: (Random seed: 42).

canvas := RSCanvas new.
shapes := RSCircle models: nodes.
canvas addAll: shapes.
RSNormalizer size
    shapes: shapes;
    normalize.

RSEquidistantCircleLayout on: shapes.
canvas @ RSCanvasController.
canvas open
```

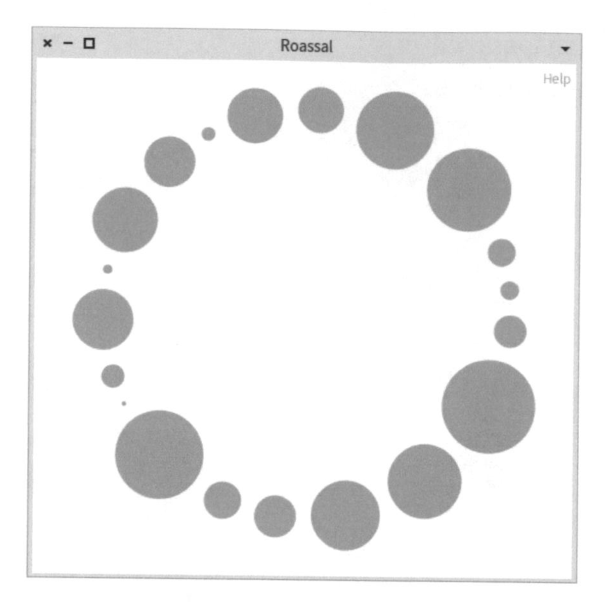

Figure 12-3. *Using the equidistant circle layout*

The equidistant circle layout tries to avoid overlapping shapes.

Grid Layout

Shapes can be located as a grid using `RSGridLayout`. Consider the following example (see Figure 12-4):

```
nodes := (1 to: 20) asArray shuffleBy: (Random seed: 42).

canvas := RSCanvas new.
shapes := RSCircle models: nodes.
canvas addAll: shapes.
RSNormalizer size
    shapes: shapes;
    normalize.

RSGridLayout on: shapes.
canvas @ RSCanvasController.
canvas open
```

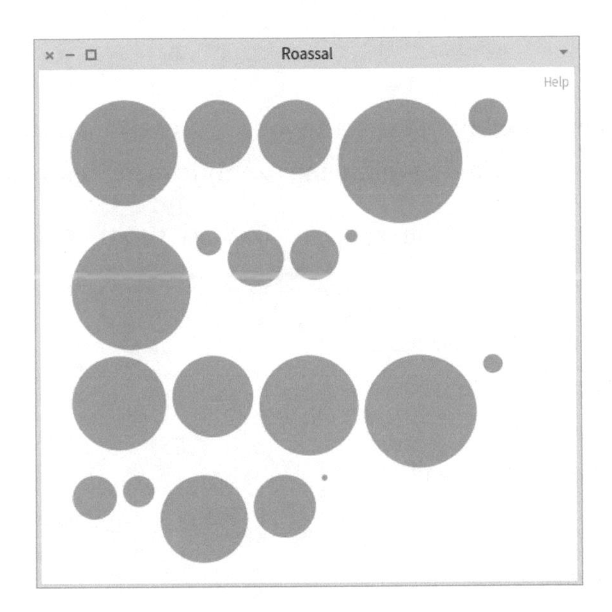

Figure 12-4. *Grid layout*

Each row of shapes has the same number of shapes, which is determined to have an overall horizontal rectangular layout (i.e., the width is larger than the height). The overall layout is determined using the *golden ratio*. This golden ratio is said to provide "pleasing, harmonious proportions" (`https://en.wikipedia.org/wiki/Golden_ratio`).

You can override the number of shapes per line by using `lineItemsCount:`, as in:

```
nodes := (1 to: 20) asArray shuffleBy: (Random seed: 42).

canvas := RSCanvas new.
shapes := RSCircle models: nodes.
canvas addAll: shapes.
RSNormalizer size
    shapes: shapes;
    normalize.

RSGridLayout new lineItemsCount: 9; on: shapes.
canvas @ RSCanvasController.
canvas open
```

Specifying an arbitrary number of shapes overrides the golden ratio proportion. The previous example forces the layout to locate nine shapes per row.

Flow Layout

The flow layout, implemented with `RSFlowLayout`, tries to fill a provided space, without reordering the shapes. Consider the following example (see Figure 12-5):

```
nodes := (1 to: 20) asArray shuffleBy: (Random seed: 42).

canvas := RSCanvas new.
shapes := RSCircle models: nodes.
canvas addAll: shapes.
RSNormalizer size
    shapes: shapes;
    normalize.

RSFlowLayout on: shapes.
canvas zoomToFit.
canvas open
```

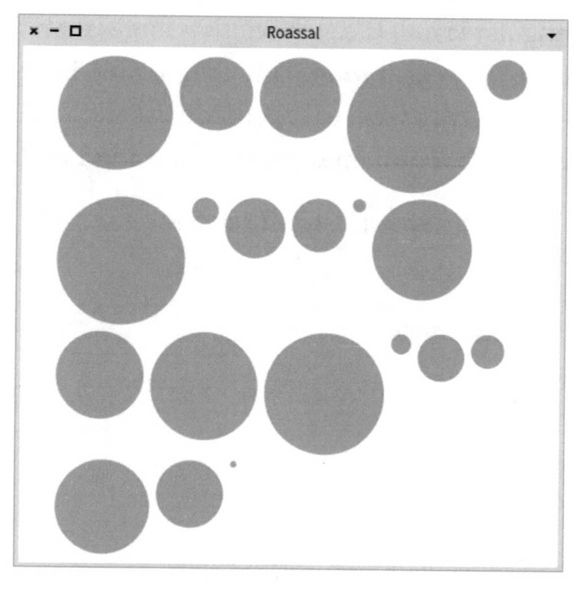

Figure 12-5. *Using the flow layout*

The RSFlowLayout class fills a given area with a constraint set on the width, using the sequence of the provided shape. A nice use of this layout is to set a callback on the canvas to automatically redo the positioning of the shapes. Consider this example:

```
nodes := (1 to: 20) asArray shuffleBy: (Random seed: 42).

canvas := RSCanvas new.
shapes := RSCircle models: nodes.
canvas addAll: shapes.
RSNormalizer size
    shapes: shapes;
    normalize.

canvas when: RSExtentChangedEvent do: [ :event |
    RSFlowLayout new
        maxWidth: event newExtent x;
        on: shapes.
    canvas zoomToFit.
].
canvas open
```

The layout is performed each time the window is resized. The width size used by the layout is set using maxWidth:. This feature can be handy when vertical scrolling is preferred over horizontal scrolling.

Rectangle Pack Layout

The rectangle pack layout, implemented with the RSRectanglePackLayout class, fills a space along its two dimensions, without preserving the initial order of the shape. Consider this example (see Figure 12-6):

```
nodes := (1 to: 20) asArray shuffleBy: (Random seed: 42).

canvas := RSCanvas new.
shapes := RSCircle models: nodes.
canvas addAll: shapes.
```

```
RSNormalizer size
    shapes: shapes;
    normalize.

RSRectanglePackLayout on: shapes.
canvas zoomToFit.
canvas open
```

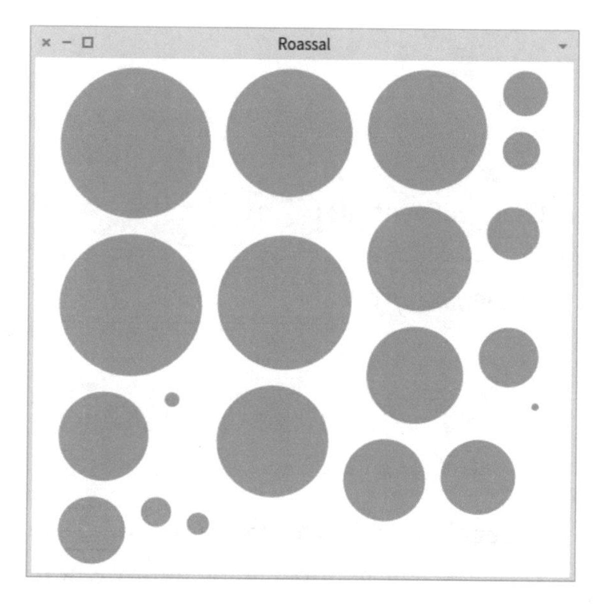

Figure 12-6. *Using the rectangle pack layout*

A larger example using the rectangle pack layout could be as follows (see Figure 12-7):

```
numberOfBoxes := 900.
r := Random seed: 42.
shapes := RSGroup new.
numberOfBoxes timesRepeat: [
    shapes add: (RSBox new width: (r nextInt: 40); height: (r nextInt: 40)) ].

canvas := RSCanvas new.
canvas addAll: shapes.
RSRectanglePackLayout new on: shapes.
canvas @ RSCanvasController.
canvas open
```

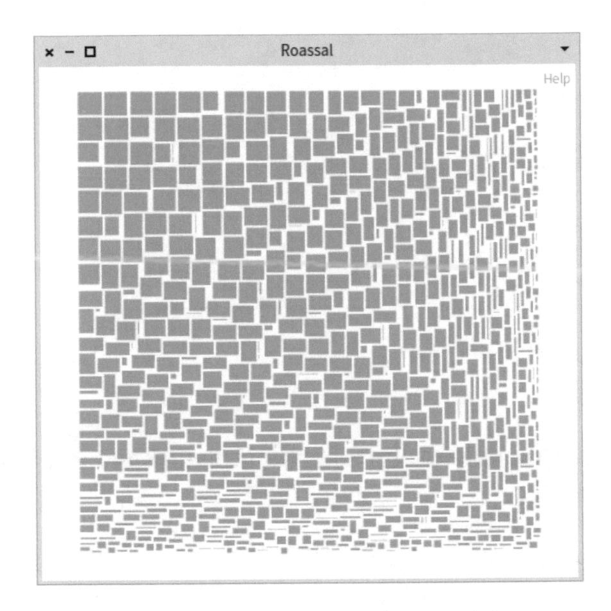

Figure 12-7. *Larger example using the rectangle pack layout*

The algorithm used by the layout tries to maximize the occupation of a bounding box. Note that the rectangle pack layout does not preserve the order of the provided shapes.

Line Layout

The `RSHorizontalLineLayout` and `RSVerticalLineLayout` classes implement line layouts. Such a layout simply consists of lining up shapes, either horizontally or vertically. Consider this example (see Figure 12-8):

```
nodes := (1 to: 20) asArray shuffleBy: (Random seed: 42).

canvas := RSCanvas new.
shapes := RSCircle models: nodes.
canvas addAll: shapes.
RSNormalizer size
    shapes: shapes;
    normalize.

RSHorizontalLineLayout on: shapes.
canvas zoomToFit.
canvas open
```

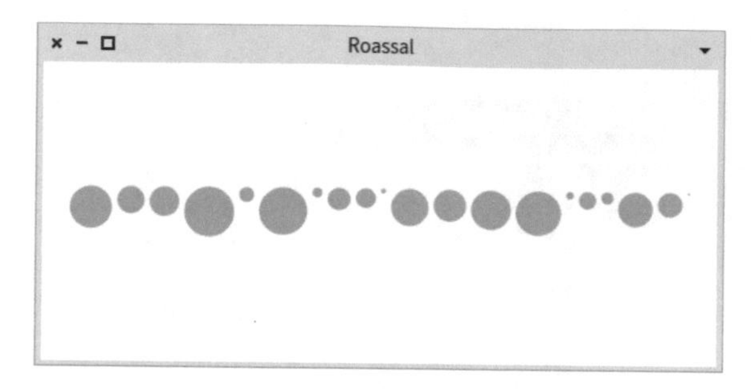

Figure 12-8. *Horizontal line layout*

The line layout is particularly useful when composing shapes. Consider the following example (see Figure 12-9):

```
canvas := RSCanvas new.

1 to: 10 do: [ :nb |
    group := RSGroup new.
    group add: (RSLabel text: nb).
    nb timesRepeat: [ group add: (RSCircle new) ].
    RSHorizontalLineLayout new alignMiddle; on: group.
    canvas add: group asShape.
].
RSVerticalLineLayout on: canvas nodes.
canvas zoomToFit.
canvas open
```

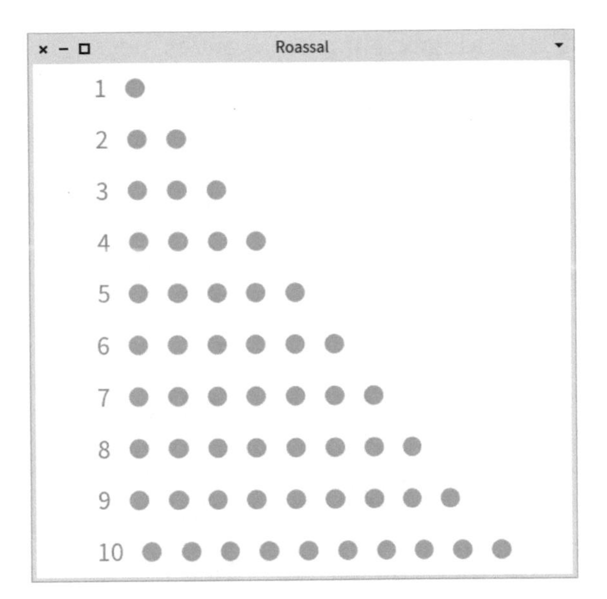

***Figure 12-9.** Line layout used with composite shapes*

The example uses the two line layouts. `RSHorizontalLineLayout` is used to horizontally line up the labels with the little circles. `RSVerticalLineLayout` is used to vertically locate the composite shapes.

These two line layouts offer handy methods to control the alignment of the shapes:

- `RSHorizontalLineLayout` provides `alignBottom`, `alignMiddle`, and `alignTop`. In addition, the alignment `alignLabel` is used to align labels along their baselines.
- `RSVerticalLineLayout` provides `alignCenter`, `alignLeft`, and `alignRight`.

Consider a slight modification of the previous script (see Figure 12-10):

```
canvas := RSCanvas new.

1 to: 10 do: [ :nb |
    group := RSGroup new.
    group add: (RSLabel text: nb).
    nb timesRepeat: [ group add: (RSCircle new) ].
    RSHorizontalLineLayout new alignMiddle; on: group.
    canvas add: group asShape.
].
```

```
RSVerticalLineLayout new alignRight; on: canvas nodes.
canvas zoomToFit.
canvas open
```

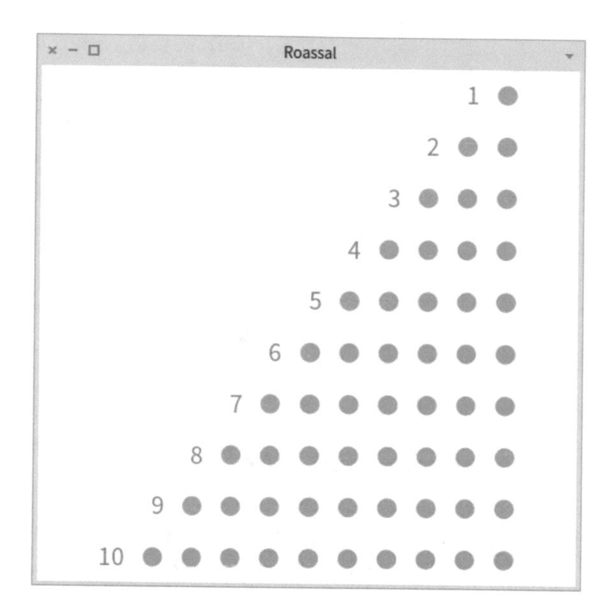

***Figure 12-10.** Use of alignment with line layouts*

The `alignRight` class used with the vertical line layout locates all the composite shapes on the right side.

Tree Layout

The layouts presented previously locate shapes without considering connections with other shapes. A number of line-driven layouts take connections into account to locate shapes. Line-driven layouts are modeled with subclasses of `RSLineDrivenLayout`.

The tree layout is a classical layout that renders a structure of the shapes as a tree. Consider this example (see Figure 12-11):

```
nodes := (1 to: 20) asArray shuffleBy: (Random seed: 42).

canvas := RSCanvas new.
shapes := RSCircle models: nodes.
shapes withBorder.
```

```
shapes color: Color white.
shapes @ RSLabeled middle.
shapes @ RSDraggable.
canvas addAll: shapes.

RSNormalizer size
    shapes: shapes;
    from: 20; to: 40;
    normalize.

RSLineBuilder line
    withBorderAttachPoint;
    shapes: shapes;
    connectFrom: [ :nb | nb // 2 ].

RSTreeLayout on: shapes.

canvas zoomToFit.
canvas open
```

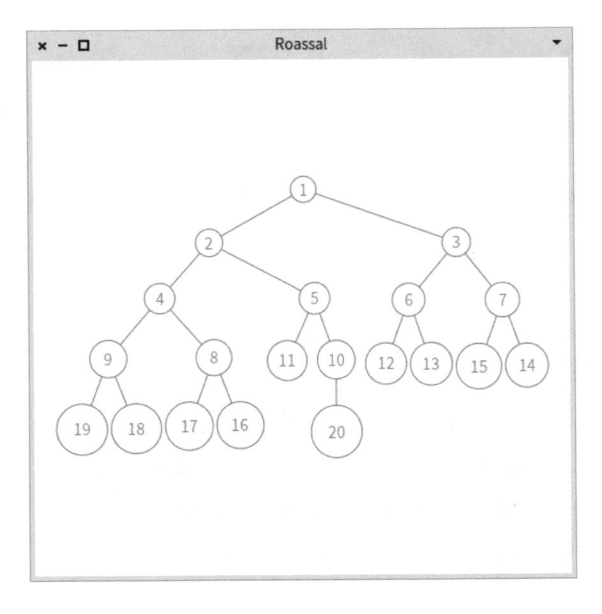

Figure 12-11. *Tree layout*

This code defines 20 circle shapes, each having a particular size and each connected to other shapes. Note that the tree layout assumes that there is no cycle between the shapes and the connections.

Another example with a less regular structure is a class hierarchy (see Figure 12-12):

```
classes := Collection withAllSubclasses.

canvas := RSCanvas new.
shapes := RSCircle models: classes.
canvas addAll: shapes.
RSLineBuilder line
    shapes: shapes;
    connectFrom: #superclass.
RSTreeLayout on: shapes.
canvas zoomToFit.
canvas open
```

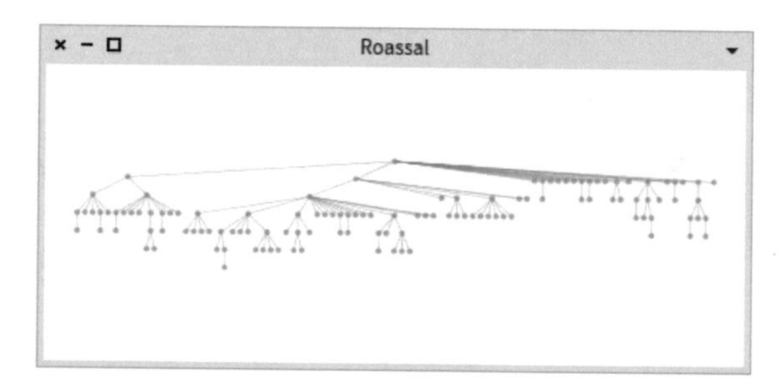

Figure 12-12. *Tree layout on a class hierarchy*

Force-Based Layout

The force-based layout (also commonly called the spring layout) is an algorithm in which nodes represent repulsing magnets and edges attracting springs. The layout uses a physical engine that often produces visually appealing results.

Consider the following example (see Figure 12-13):

```
numberOfNodes := 25.
numberOfLines := 40.
r := Random seed: 42.
graph := Dictionary new.
```

```
1 to: numberOfNodes do: [ :aNode |
    graph at: aNode put: Set new ].

numberOfLines timesRepeat: [
    fromNode := r nextInteger: numberOfNodes.
    toNode := r nextInteger: numberOfNodes.
    (graph at: fromNode) add: toNode ].

canvas := RSCanvas new.
nodes := RSCircle models: (1 to: numberOfNodes).
nodes color: #red.
nodes @ RSDraggable @ RSPopup.
canvas addAll: nodes.

lb := RSLineBuilder line.
lb canvas: canvas.
lb withBorderAttachPoint.
lb makeBidirectional.
lb moveBehind.
lb objects: (1 to: numberOfNodes).
lb connectToAll: [ :aNumber | graph at: aNumber ].

RSForceBasedLayout new charge: -300; on: nodes.
canvas @ RSCanvasController.
canvas open
```

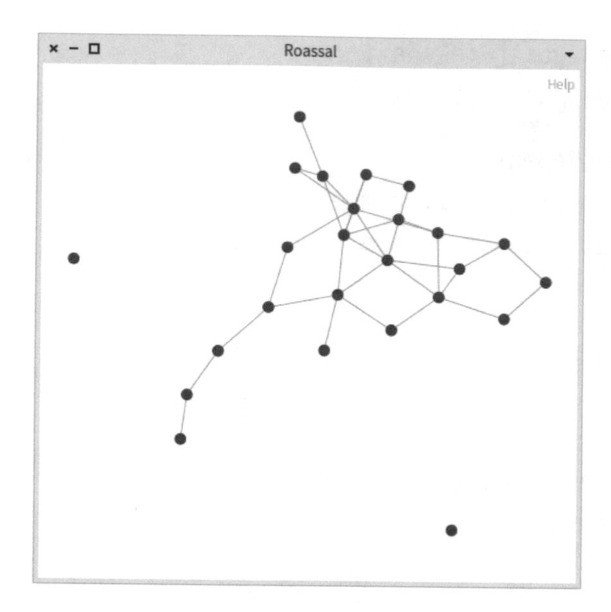

Figure 12-13. *Force-based layout*

The result of the graph layout is organic and often pleasant to see. Some parameters help shape the behavior of the layout. In particular:

- `charge:` Takes a negative number that represents the negative charge of the nodes.
- `friction:` Takes a number between 0 and 1 and represents the friction when nodes are moved around. A high friction makes the nodes move slowly.
- `length:` Sets the length of the edges and accepts a positive number.
- `iterations:` Sets an arbitrary number of iterations.

From these four parameters, only `charge` is set in the previous examples. The three remaining parameters have a default value that suits the example.

Conditional Layout

Layouts often need to be composed according to the visual layout you want to obtain. Consider the following example (see Figure 12-14):

```
numberOfNodes := 50.
```

```
numberOfLines := 20.
r := Random seed: 42.
graph := Dictionary new.

1 to: numberOfNodes do: [ :aNode |
    graph at: aNode put: Set new ].

numberOfLines timesRepeat: [
    fromNode := r nextInteger: numberOfNodes.
    toNode := r nextInteger: numberOfNodes.
    (graph at: fromNode) add: toNode ].

canvas := RSCanvas new.
nodes := RSCircle models: (1 to: numberOfNodes).
nodes color: #red.
nodes @ RSDraggable @ RSPopup.
canvas addAll: nodes.

lb := RSLineBuilder line.
lb canvas: canvas.
lb withBorderAttachPoint.
lb makeBidirectional.
lb moveBehind.
lb objects: (1 to: numberOfNodes).
lb connectToAll: [ :aNumber | graph at: aNumber ].

RSForceBasedLayout new charge: -100; on: nodes.
canvas @ RSCanvasController.
canvas open
```

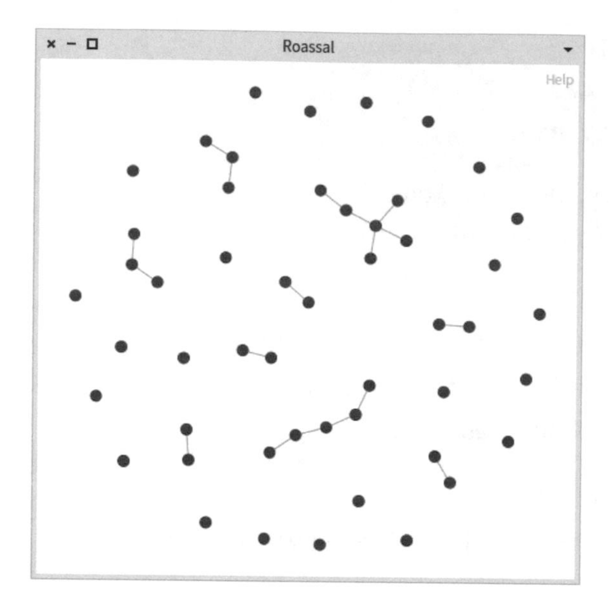

***Figure 12-14.** The force-based layout on a sparse graph*

The script builds and visualizes a graph made of 50 nodes and 20 edges. As Figure 12-14 illustrates, the graph is sparsely connected, as it contains many more nodes than edges. An easy way to order this visualization is to move all non-connected nodes to one side. Some conditions can be set on a structure to reflect particular decisions that ultimately aim at improving the effectiveness of the graph. Consider the following improvement (see Figure 12-15):

```
numberOfNodes := 50.
numberOfLines := 20.
r := Random seed: 42.
graph := Dictionary new.

1 to: numberOfNodes do: [ :aNode |
    graph at: aNode put: Set new ].

numberOfLines timesRepeat: [
    fromNode := r nextInteger: numberOfNodes.
    toNode := r nextInteger: numberOfNodes.
    (graph at: fromNode) add: toNode ].
```

```
canvas := RSCanvas new.
nodes := RSCircle models: (1 to: numberOfNodes).
nodes color: #red.
nodes @ RSDraggable @ RSPopup.
canvas addAll: nodes.

lb := RSLineBuilder line.
lb canvas: canvas.
lb withBorderAttachPoint.
lb makeBidirectional.
lb moveBehind.
lb objects: (1 to: numberOfNodes).
lb connectToAll: [ :aNumber | graph at: aNumber ].

RSConditionalLayout new
    ifNotConnectedThen: RSGridLayout new;
    else: (RSForceBasedLayout new charge: -300);
    on: nodes.
canvas @ RSCanvasController.
canvas open
```

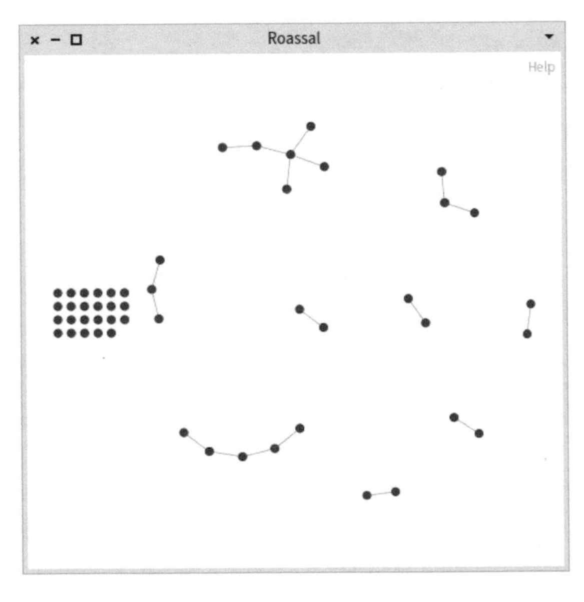

Figure 12-15. *A grid layout is applied to unconnected nodes while connected nodes use a force-based layout*

The `RSConditionalLayout` class is a conditional layout. The class defines a method called `if:` that accepts a one-argument block evaluated on each node. The layout provided to `then:` is applied to the nodes that meet the condition. The layout provided to `else:` is applied to the nodes that do not meet the condition. Utility methods are provided to ease the writing. In particular:

- `ifConnectedThen: aLayout` applies `aLayout` to the nodes that are connected.
- `ifNotConnectedThen: aLayout` applies `aLayout` to the nodes that are not connected.

The previous example applies a grid layout to the orphan nodes, while a force layout is applied to the connected ones. However, each cluster can be separately handled to reduce spaces between connected nodes. For example (see Figure 12-16):

```
numberOfNodes := 50.
numberOfLines := 20.
r := Random seed: 42.
graph := Dictionary new.

1 to: numberOfNodes do: [ :aNode |
    graph at: aNode put: Set new ].

numberOfLines timesRepeat: [
    fromNode := r nextInteger: numberOfNodes.
    toNode := r nextInteger: numberOfNodes.
    (graph at: fromNode) add: toNode ].

canvas := RSCanvas new.
nodes := RSCircle models: (1 to: numberOfNodes).
nodes color: #red.
nodes @ RSDraggable @ RSPopup.
canvas addAll: nodes.

lb := RSLineBuilder line.
lb canvas: canvas.
lb withBorderAttachPoint.
lb makeBidirectional.
```

```
lb moveBehind.
lb objects: (1 to: numberOfNodes).
lb connectToAll: [ :aNumber | graph at: aNumber ].

RSConditionalLayout new
    ifNotConnectedThen: RSGridLayout new;
    else: (RSClusteringLayout new
                clustersLayout: RSFlowLayout new;
                   forEachLayout: (RSForceBasedLayout new charge: -300));
    on: nodes.
canvas @ RSCanvasController.
canvas open
```

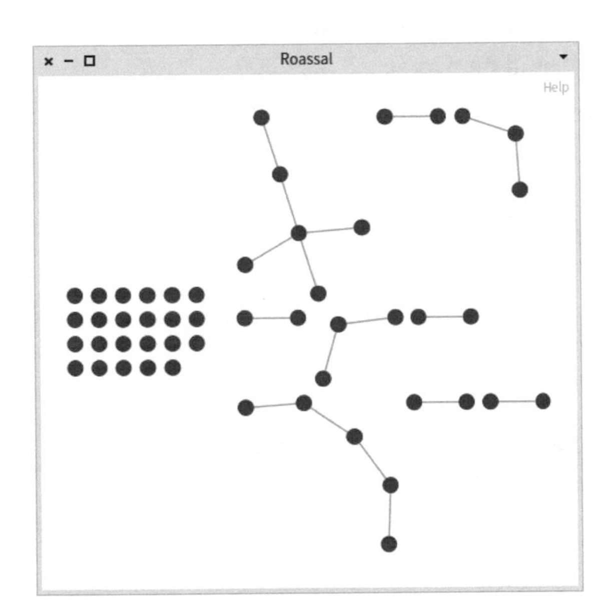

***Figure 12-16.** Combining conditional and clustering layouts*

The `RSClusteringLayout` class is a layout that identifies clusters of connected nodes and applies a layout to each group. In the previous example, the force layout is applied to each cluster and the flow layout is applied to all the clusters. Structuring a layout that way is convenient to reduce the amount of space taken by the graph.

Graphviz Layouts

Graphviz is an open source graph visualization library that is well supported by an international community. Roassal can use Graphviz in a transparent way to perform a graph layout. Graphviz is highly optimized and can perform a layout on a large graph (e.g., a couple thousands of nodes and edges).

Installing Graphviz

The first step to using Graphviz is to install it on your computer. Installing Graphviz is independent of Pharo. The official website of Graphviz is `https://graphviz.org` and the `https://graphviz.org/download/` page provides instructions on how to install Graphviz on your favorite operating system. For Windows, the download page provides links to download the install packages.

On the Ubuntu distribution of Linux, the following instructions should install Graphviz when executed in a terminal:

```
sudo apt update
sudo apt install graphviz
```

On macOS, you can use `brew` from the command line:

```
brew install graphviz
```

If you do not have `brew` available on macOS, you can install it easily. The `https://brew.sh` website gives instructions on how to install `brew`.

At the time of writing this chapter, there is no distribution of Graphviz for an M1-based macOS machine (ARM). You can execute `arch -x86_64 brew install graphviz`, which should install the x86 version of Graphviz, if you have an M1 machine.

You can verify your installation of Graphviz by executing the following instruction in a terminal:

```
echo "digraph G {Hello->World}" | dot -Tpng > hello.png
```

This example should create a picture entitled `hello.png` that contains two connected nodes. If it does not, something went wrong with the installation. If the installation did not lead to an error, the usual suspect is making sure that the executable dot is accessible by setting the `PATH` shell variable. If the `hello.png` file cannot be created, the remaining of the chapter will not work.

Bridging Roassal and Graphviz

After installing Graphviz on your machine, you can install the Roassal/Graphviz bridge by executing the following instruction:

```
[ Metacello new
    baseline: 'Roassal3';
    repository: 'github://ObjectProfile/Roassal3';
    load: 'Graphviz' ] on: MCMergeOrLoadWarning do: [ :warning | warning
    load ].
```

The `RSGraphvizLayout` class is the entry point to using the Graphviz bridge.

Graphviz Layout

Consider the following example (see Figure 12-17):

```
c := RSCanvas new.

shapes := RSLabel models: ByteArray withAllSubclasses.
c addAll: shapes.
shapes @ RSDraggable.

lb := RSLineBuilder graphviz.
lb shapes: shapes.
lb connectFrom: #superclass.

RSGraphvizLayout on: shapes.
c @ RSCanvasController.
c open
```

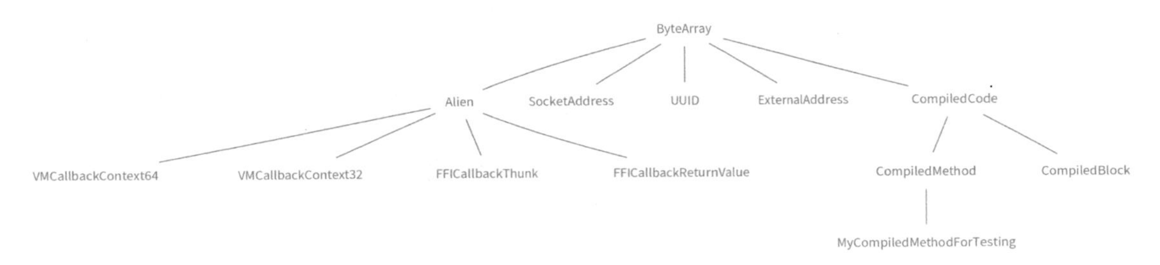

Figure 12-17. *Visualizing classes*

Figure 12-17 shows the hierarchy of classes that use the Graphviz layout. As a more complex example, consider this script (see Figure 12-18):

```
c := RSCanvas new.

shapes := RSLabel models: Collection withAllSubclasses.
c addAll: shapes.
shapes @ RSDraggable.

lb := RSLineBuilder graphviz.
lb arrowMarkerEnd.
lb shapes: shapes.
lb connectToAll: #dependentClasses.

RSGraphVizLayout on: shapes.
c @ RSCanvasController.
c open
```

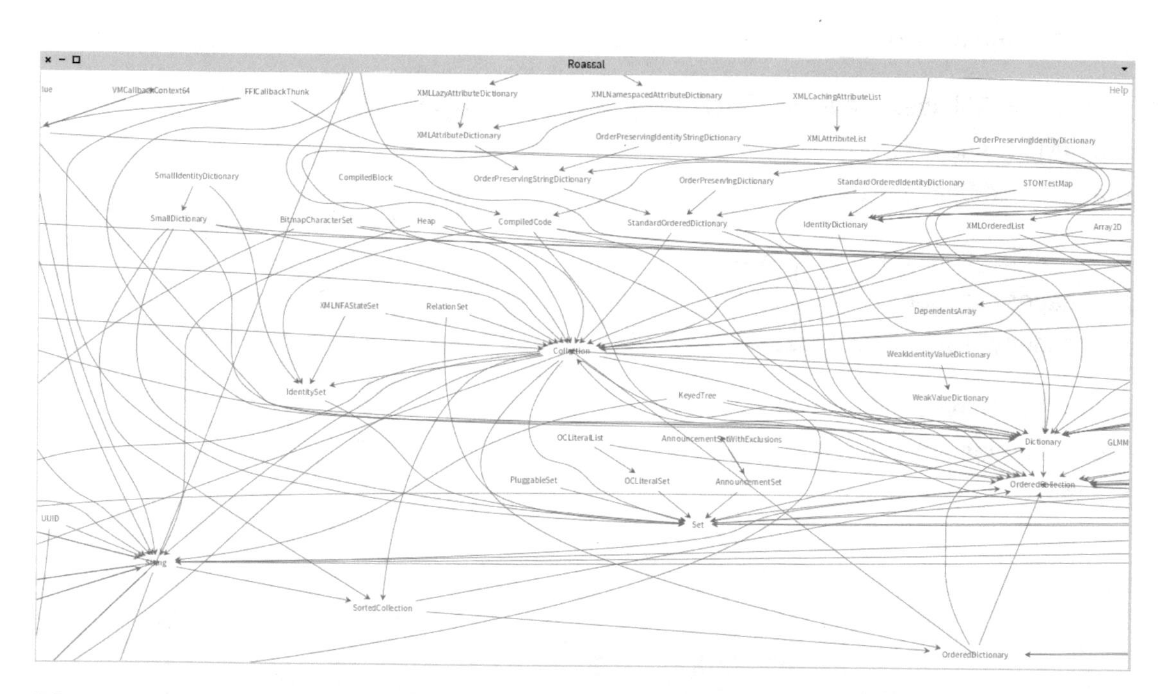

Figure 12-18. Complex graph layout using Graphviz (detail)

Graphviz offers numerous options to tune a layout. I recommend you explore these possibilities with the `RSGraphVizExamples` class.

What Have You Learned in This Chapter?

Applying a layout to a graph is an essential ability of Roassal. A number of layouts are provided to accommodate most of the common needs when visualizing data. In particular, the chapter detailed the following:

- Discussed layouts that do not explore the graph structure (e.g., circle, grid, flow, rectangle pack, and line).
- Discussed a number of line-driven layouts (e.g., tree and force).
- Briefly outlined the ability of Roassal to use Graphviz as a layout.

CHAPTER 13

Integration in the Inspector

Pharo offers many powerful and extensible software development tools. One of them is the Inspector, designed to inspect any Pharo object. The Inspector is a central tool in Pharo as it allows you to see the internal representation of an object, intimately interact with an object through an evaluation panel, and define visual representations of an object.

Visualizations built with Roassal can be hooked into the Inspector. Furthermore, navigation through the graph of objects happens by simply clicking Roassal shapes, which can open a new visualization, itself showing clickable shapes. This chapter details the mechanism to embed Roassal visualizations in the Pharo Inspector.

In addition to detailing the support of Roassal by the Inspector, the chapter illustrates the use of Chart (the charting library of Roassal) and Mondrian (the graph library of Roassal).

Pharo Inspector

The Pharo Inspector can be invoked in a number of ways. You can send the `inspect` message to any object to open the Inspector. For example, you can open the Playground and execute the following code by pressing Cmd+D (on macOS) or Alt+D (on Linux and Windows) or by right-clicking Do It (see Figure 13-1):

```
c := OrderedCollection new.
c add: 10.
c add: 15.
c add: 19.
c inspect
```


A. Bergel, *Agile Visualization with Pharo*, https://doi.org/10.1007/978-1-4842-7161-2_13

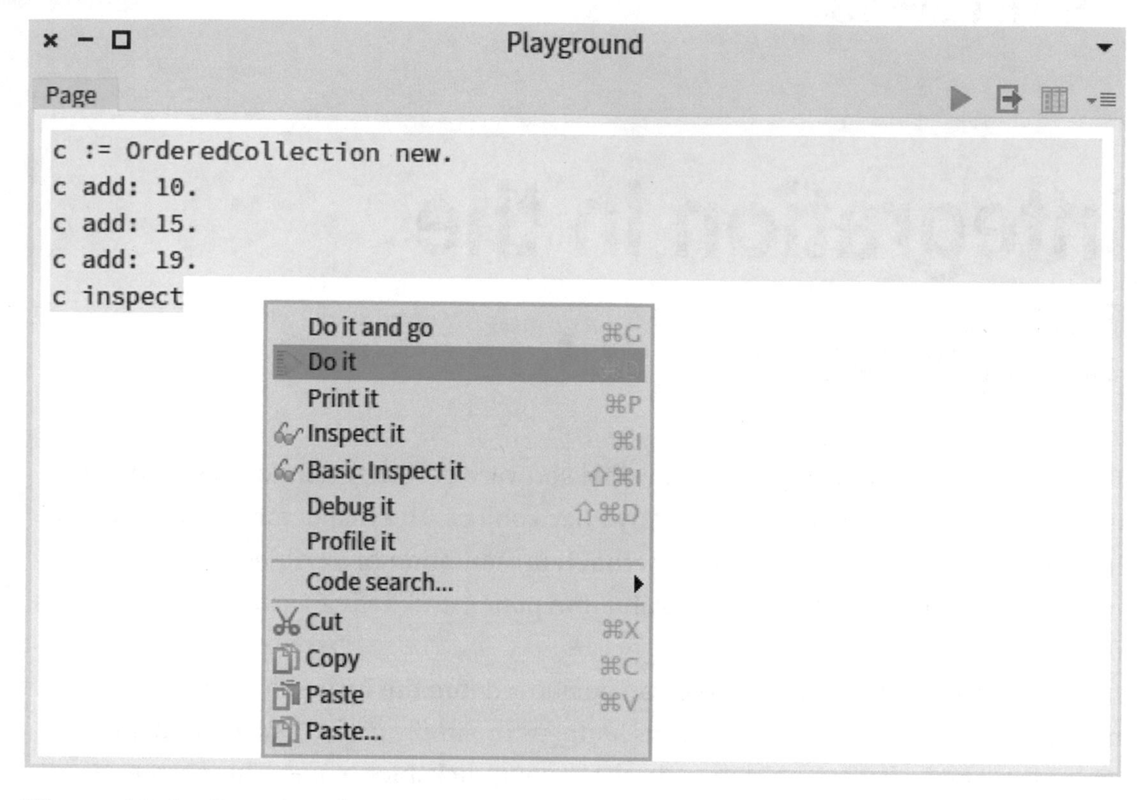

Figure 13-1. *Opening the Inspector*

The action of sending `inspect` opens a new window. Although this can be convenient in many situations, the Inspector can also be open within the Playground window. You can press Cmd+G/Alt+G to embed the Inspector in the Playground window using the following code. In such a case, the `inspect` message is not necessary (see Figure 13-2):

```
c := OrderedCollection new.
c add: 10.
c add: 15.
c add: 19.
c
```

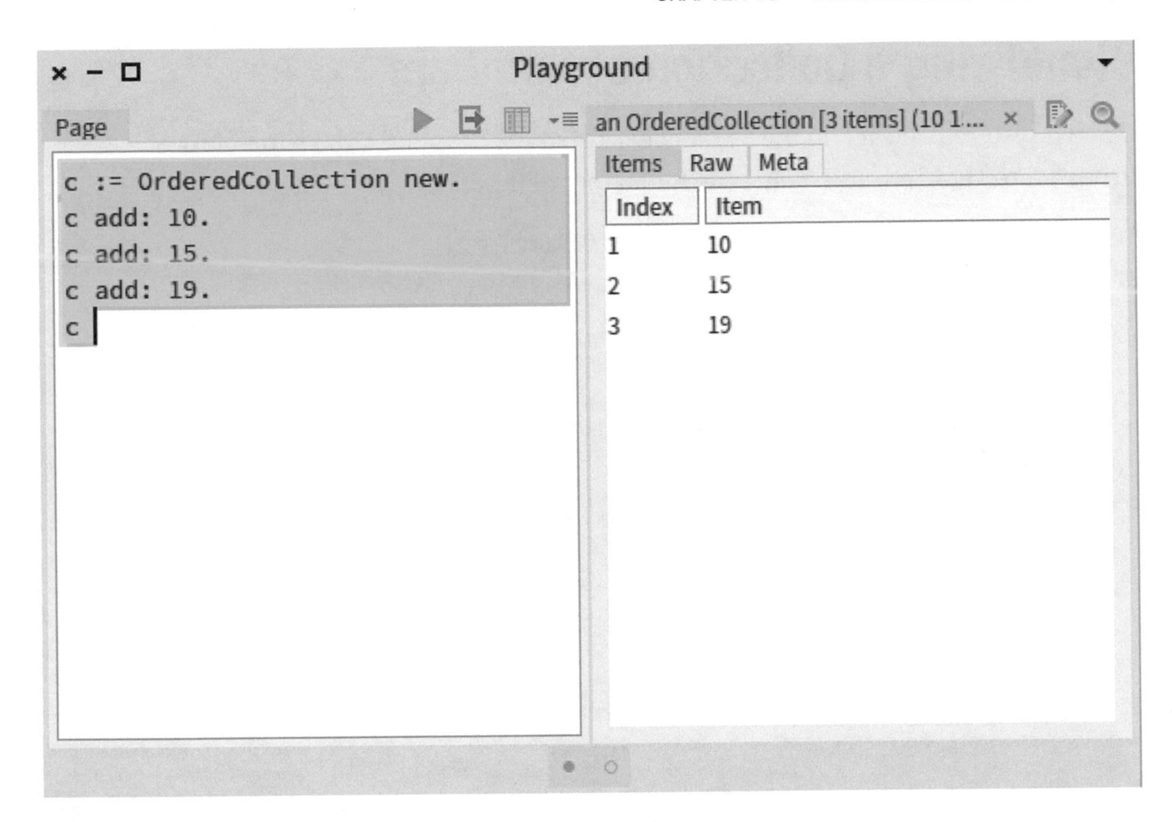

Figure 13-2. *Inspecting a collection*

Invoking the Cmd+G/Alt+G keystroke splits the Playground into two panes, with the Playground on the left and the Inspector on the right. The Inspector has three tabs—*Items*, *Raw*, and *Meta*. The first tab (Items) lists the items contained in the collection. The second tab (Raw) shows the internal values of the object representing the ordered collection. The third tab (Meta) shows the list of methods defined in the class of the object, `OrderedCollection`. The tabs that are offered by the Inspector depend on the object to be inspected. For example, inspecting a color object gives a different set of tabs.

In the following sections, you will see how to plug a Roassal visualization as a tab when inspecting an ordered collection. In particular, you will define new tabs to access visualizations.

Visualizing a Collection of Numbers

Open a system browser on the SequenceableCollection class, the superclass of OrderedCollection, and define the following method:

```
SequenceableCollection>>visualizeListOfNumbers
    | c d |
    c := RSChart new.

    d := RSLinePlot new.
    d y: self.
    c addPlot: d.
    c build.
    ^ c canvas
```

In the system browser, you should see Figure 13-3.

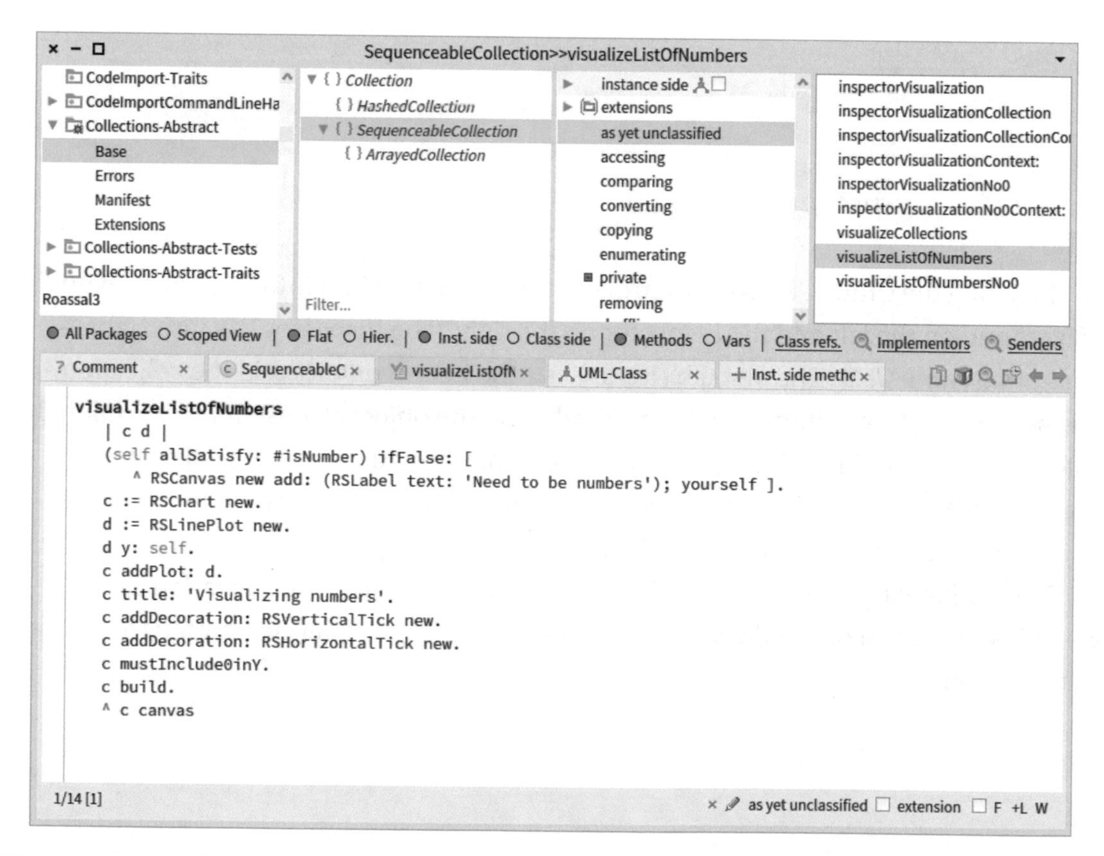

Figure 13-3. *The visualizeListOfNumbers method, together with the methods defined in this chapter*

This visualization can be invoked in the Playground using Cmd+G/Alt+G (see Figure 13-4):

```
c := OrderedCollection new.
c add: 10.
c add: 15.
c add: 19.
c visualizeListOfNumbers
```

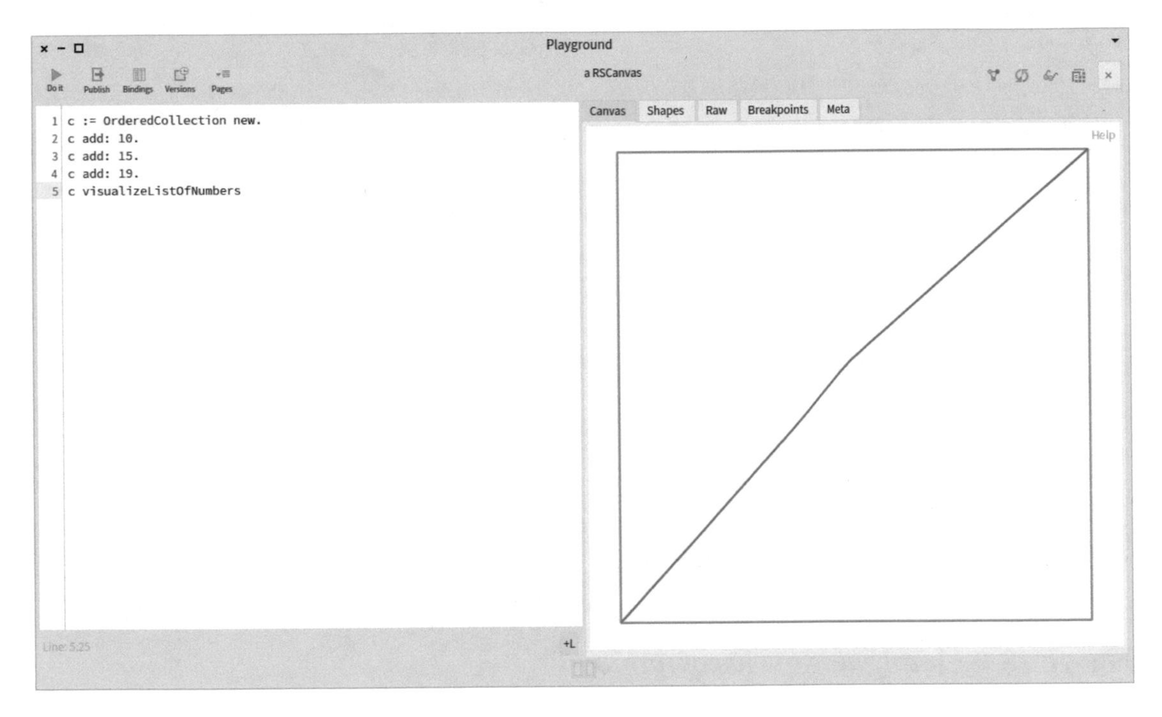

Figure 13-4. *Visualizing numbers*

The Playground splits and the visualization appears on the right pane. This visualization can be improved by adding title ticks with the following method (see Figure 13-5):

```
SequenceableCollection>>visualizeListOfNumbers
    | c d |
    c := RSChart new.
    d := RSLinePlot new.
    d y: self.
```

```
    c addPlot: d.
    c title: 'Visualizing numbers'.
    c addDecoration: RSVerticalTick new.
    c addDecoration: RSHorizontalTick new.
    c build.
    ^ c canvas
```

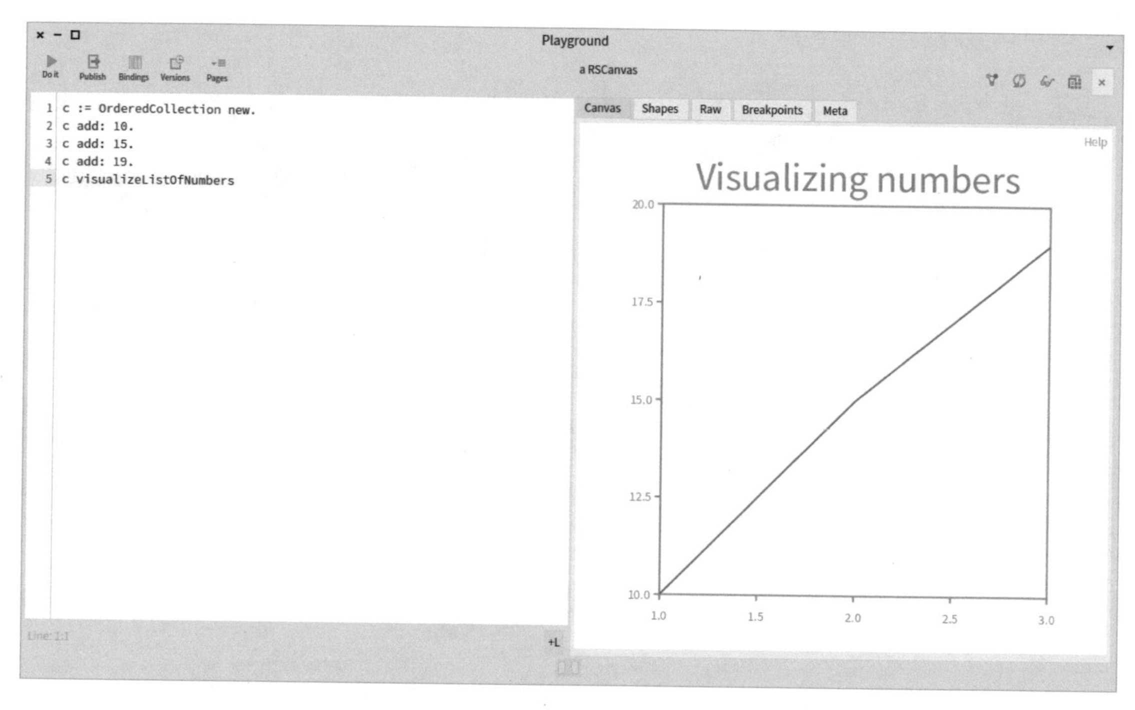

***Figure 13-5.** Improving the list of numbers*

If you want to include the value 0, use the mustInclude0inY method, as in this new version (see Figure 13-6):

```
SequenceableCollection>>visualizeListOfNumbers
    | c d |
    (self allSatisfy: #isNumber) ifFalse: [
        ^ RSCanvas new add:
            (RSLabel text: 'Need to contain numbers'); yourself ].
    c := RSChart new.
    d := RSLinePlot new.
    d y: self.
```

```
c addPlot: d.
c title: 'Visualizing numbers'.
c addDecoration: RSVerticalTick new.
c addDecoration: RSHorizontalTick new.
c mustInclude0inY.
c build.
^ c canvas
```

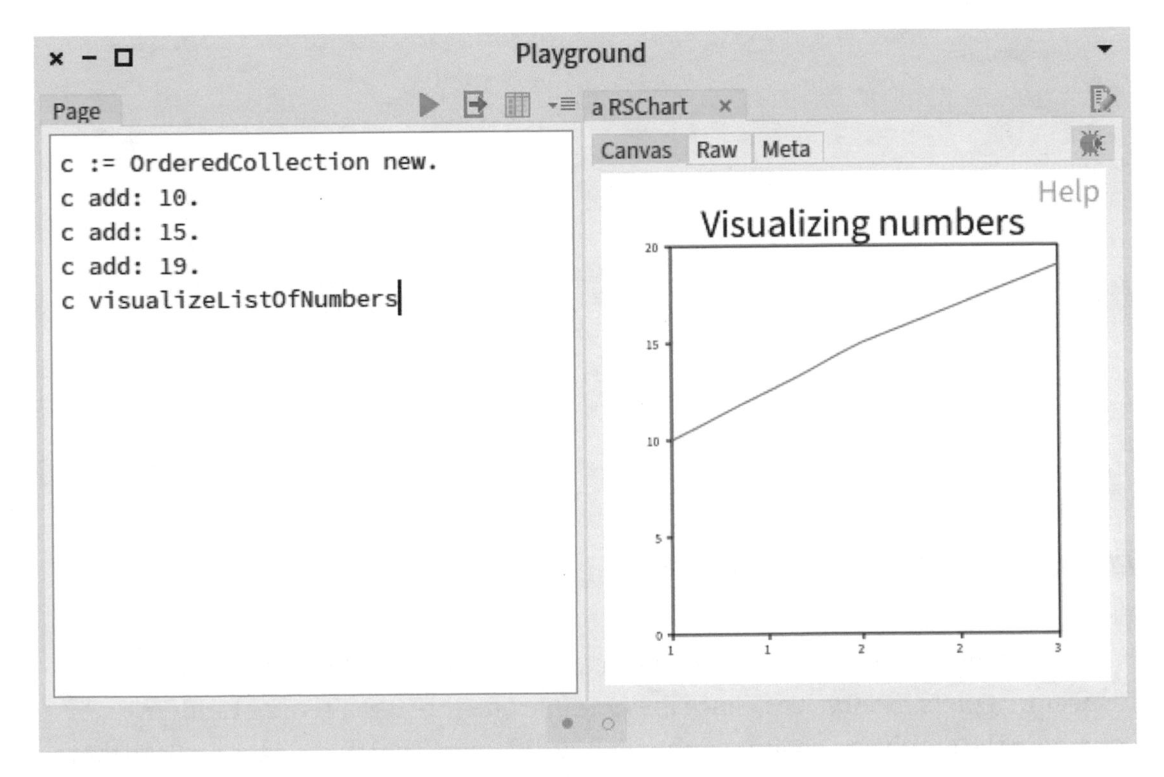

Figure 13-6. Improving the list of numbers

This new version of visualizeListOfNumbers has a guard to prevent building a chart if the collection does not contain numbers. This small visualization easily scales up. Consider the following script (see Figure 13-7):

```
numberOfValues := 1000.
y := 0.
c := OrderedCollection new.
numberOfValues timesRepeat: [
```

```
    c add: y.
    y := y + ((-30 to: 30) atRandom) ].
c visualizeListOfNumbers
```

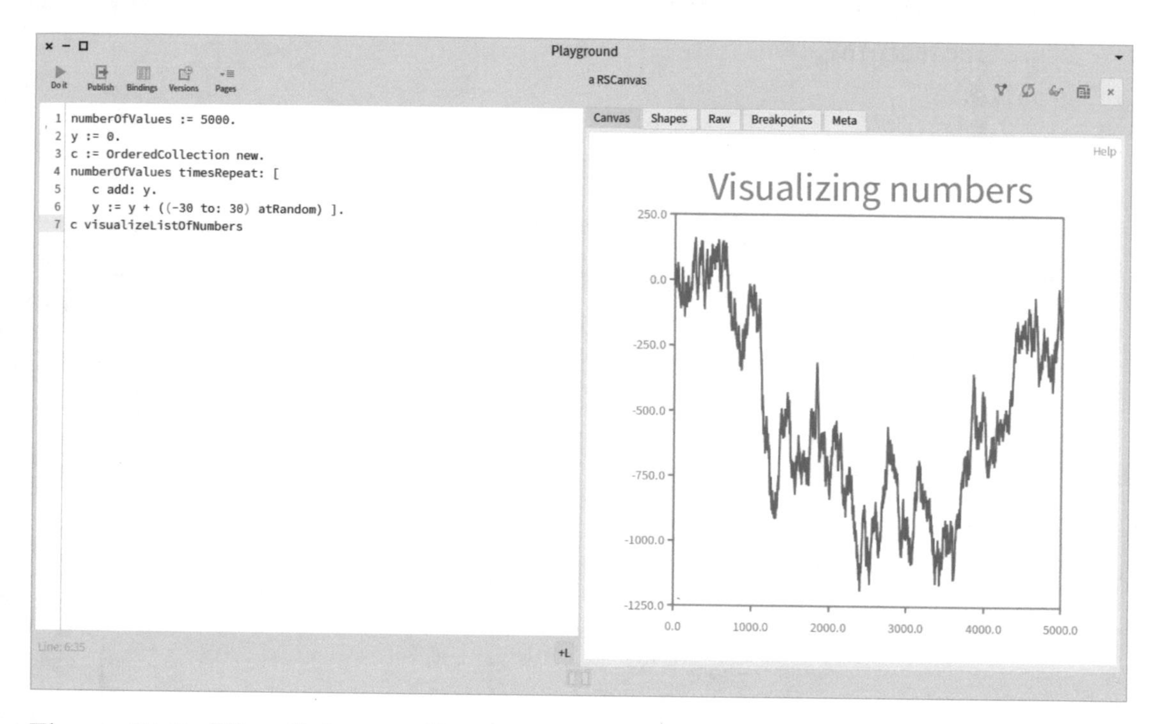

Figure 13-7. *Visualizing a collection containing numbers*

This example illustrates how a Roassal canvas can be rendered in the Pharo Inspector. It plots 1,000 values. Since the example does not use a seeded random generator, the graph you obtain may be slightly different. The `visualizeListOfNumbers` method is called by the script to produce the visualization. However, the integration of Roassal in the Inspector framework may produce the visualization by *automatically* invoking `visualizeListOfNumbers`. You can define the `inspectorVisualization` method to wrap a Roassal canvas into a `SpRoassal3InspectorPresenter` object:

```
SequenceableCollection>>inspectorVisualization
    <inspectorPresentationOrder: 90 title: 'Visualization'>

    ^ SpRoassal3InspectorPresenter new
        canvas: self visualizeListOfNumbers;
        yourself
```

The value 90 provided to the `inspectorPresentationOrder:` keyword indicates the priority of the tab in the Inspector. A low value indicates that the `Visualization` tab is located on the left. Since you do not need the evaluation pane, the method can be defined:

```
SequenceableCollection>>inspectorVisualizationContext: aContext
    aContext withoutEvaluator
```

The `inspectorVisualization` and `inspectorVisualizationContext:` methods are semantically connected by having the first as the prefix of the second.

After defining `inspectorVisualization` and `inspectorVisualizationContext:`, evaluating the following script using Cmd+G/Alt+G embeds the visualization in the Inspector (see Figure 13-8):

```
numberOfValues := 1000.
y := 0.
c := OrderedCollection new.
numberOfValues timesRepeat: [
    c add: y.
    y := y + ((-30 to: 30) atRandom) ].
c
```

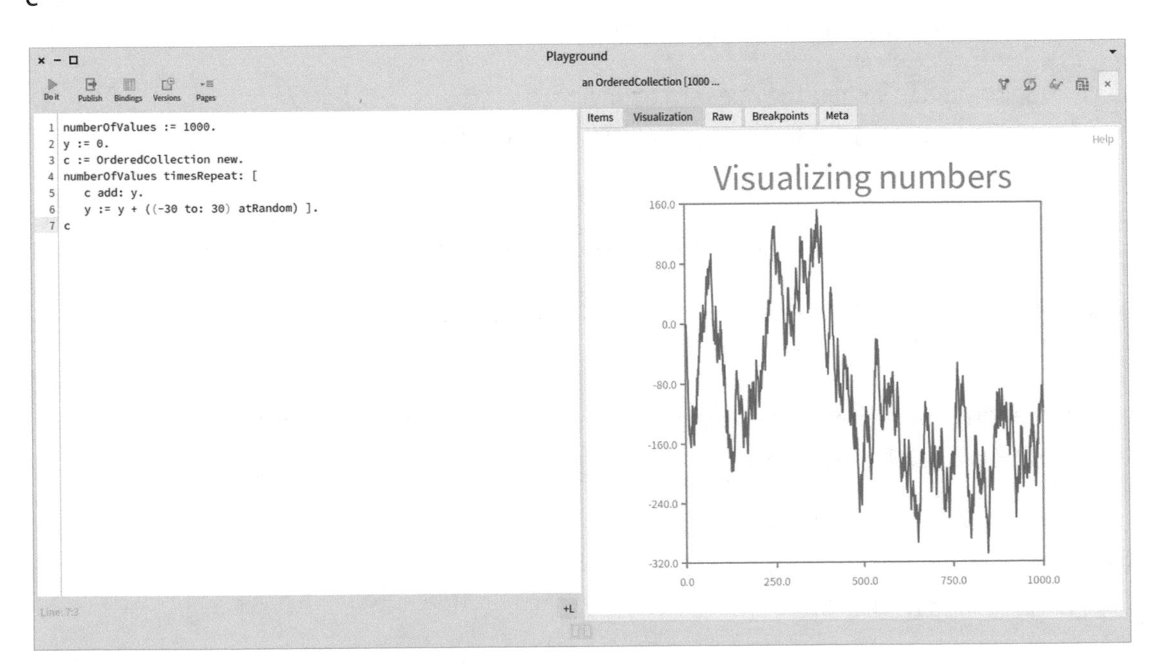

***Figure 13-8.** Embedding the visualization in the Inspector*

If you wanted to keep the chart without the 0 value in the two axes, you could define this method:

```
SequenceableCollection>>visualizeListOfNumbersNoO
    | c d |
    (self allSatisfy: #isNumber) ifFalse: [
        ^ RSCanvas new add:
            (RSLabel text: 'Need to contain numbers'); yourself ].
    c := RSChart new.
    d := RSLinePlot new.
    d y: self.
    c addPlot: d.
    c title: 'Visualizing numbers'.
    c addDecoration: RSVerticalTick new.
    c addDecoration: RSHorizontalTick new.
    c build.
    ^ c canvas
```

The canvas returned by `visualizeListOfNumbersNoO` has to be wrapped as follows:

```
SequenceableCollection>>inspectorVisualizationNoO
    <inspectorPresentationOrder: 90 title: 'VisualizationNoO'>

    ^ SpRoassal3InspectorPresenter new
        canvas: self visualizeListOfNumbersNoO;
        yourself
```

As seen previously, the evaluation pane can be suppressed using the following:

```
SequenceableCollection>>inspectorVisualizationNoOContext: aContext
    aContext withoutEvaluator
```

Inspecting the following script highlights the differences between the two visualizations (see Figure 13-9):

```
c := OrderedCollection new.
-3.14 to: 3.14 by: 0.1 do: [ :x |
    c add: x sin + 3 ].
c
```

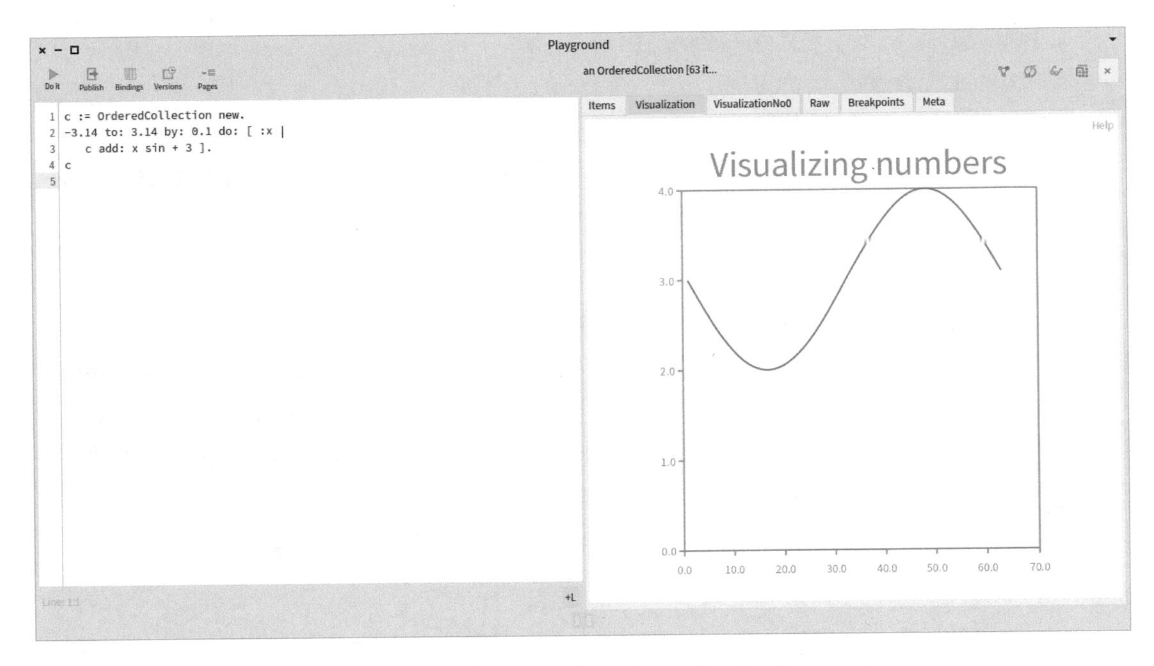

Figure 13-9. *Embedding multiple visualizations in the Inspector*

Figure 13-9 shows that the two visualizations are accessible through a dedicated tab in the Inspector. The values kept in the `c` variable range from 2 to 4. In one visualization, the origin (0, 0) is part of the graph, in the other visualization, it is not. This contrived example illustrates the ability of the Pharo Inspector to render multiple visualizations.

Chaining Visualizations

A collection may contain other collections. In this section, you see how to define a visualization that enables you to navigate through nested collections. Consider this method:

```
SequenceableCollection>>visualizeCollections
    | m |
    (self allSatisfy: #isCollection) ifFalse: [
        ^ RSCanvas new add:
            (RSLabel text: 'Need to contain collections'); yourself ].
    m := RSMondrian new.
    m shape circle color: Color gray translucent.
    m nodes: self.
```

```
    m normalizeSize: #size from: 5 to: 10.
    m layout force.
    m line connectToAll:
            [ :coll | self select: [ :coll2 | coll size = coll2 size ] ].
    m build.
    ^ m canvas
```

The visualizeCollections method makes sense only if the collection contains other collections. If not, a gentle canvas indicates an error, as expressed in the first lines of the method. The Mondrian API is used to render each collection as a transparent circle. The size of a circle reflects the number of values contained in the represented collection. Circles of the same size are connected with a line.

The visualization is hooked into the Inspector using the following:

```
SequenceableCollection>>inspectorVisualizationColl
    <inspectorPresentationOrder: 90 title: 'Collections'>

    ^ SpRoassal3InspectorPresenter new
        canvas: self visualizeCollections;
        yourself
```

Since there is no need to have an evaluator pane, you can define it as follows:

```
SequenceableCollection>>inspectorVisualizationCollContext: aContext
    aContext withoutEvaluator
```

Consider this simple example (see Figure 13-10):

```
x := (-3.14 to: 3.14 by: 0.01).
y1 := x collect: #sin.
y2 := x collect: #tan.
y3 := x collect: #cos.

{ x . y1 . y2 . y3 }
```

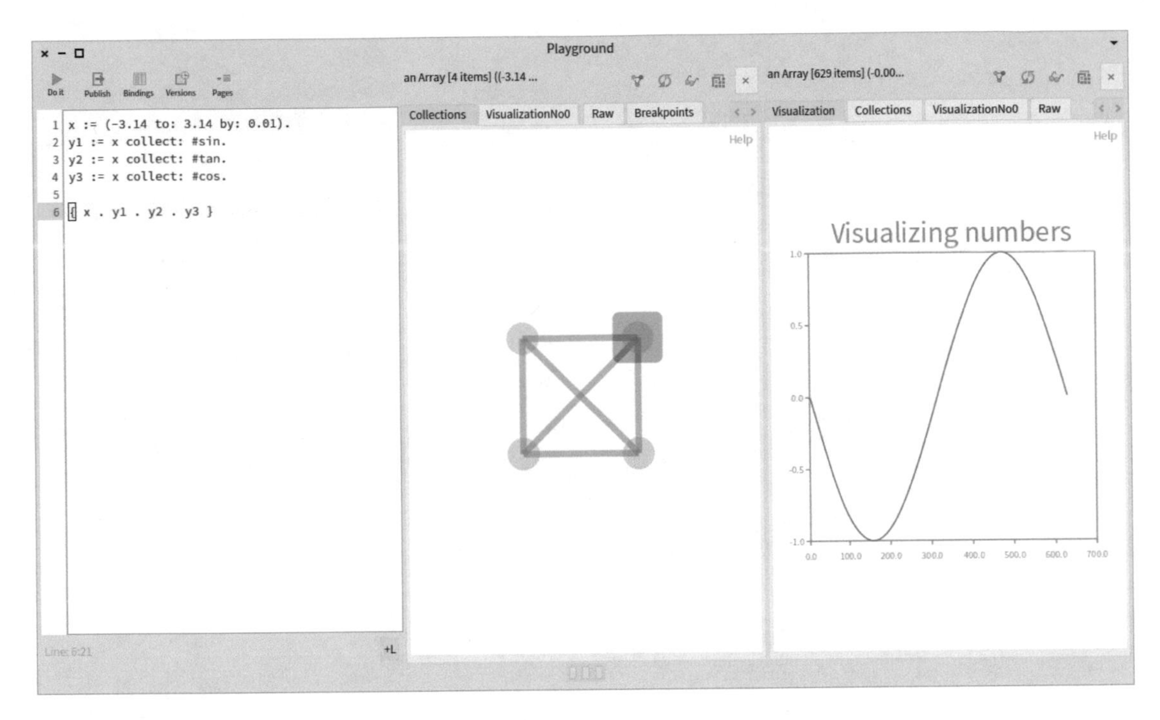

Figure 13-10. *Visualizing nested collections*

Figure 13-10 shows three panes. The middle pane is the collection that contains four other collections. Since these four collections are the same size, they are all connected to each other with a line. Clicking a node in the graph opens a new pane, which renders the visualizations you have previously seen.

In the same spirit, here is a more complex example (see Figure 13-11):

```
numberOfCollections := 100.
c := OrderedCollection new.
r := Random seed: 42.

numberOfCollections timesRepeat: [
    t := OrderedCollection new.
    y := 0.
    (r nextInteger: 100) timesRepeat: [
        t add: y.
        y := y + ((-30 to: 30) atRandom) ].
    c add: t
].

c
```

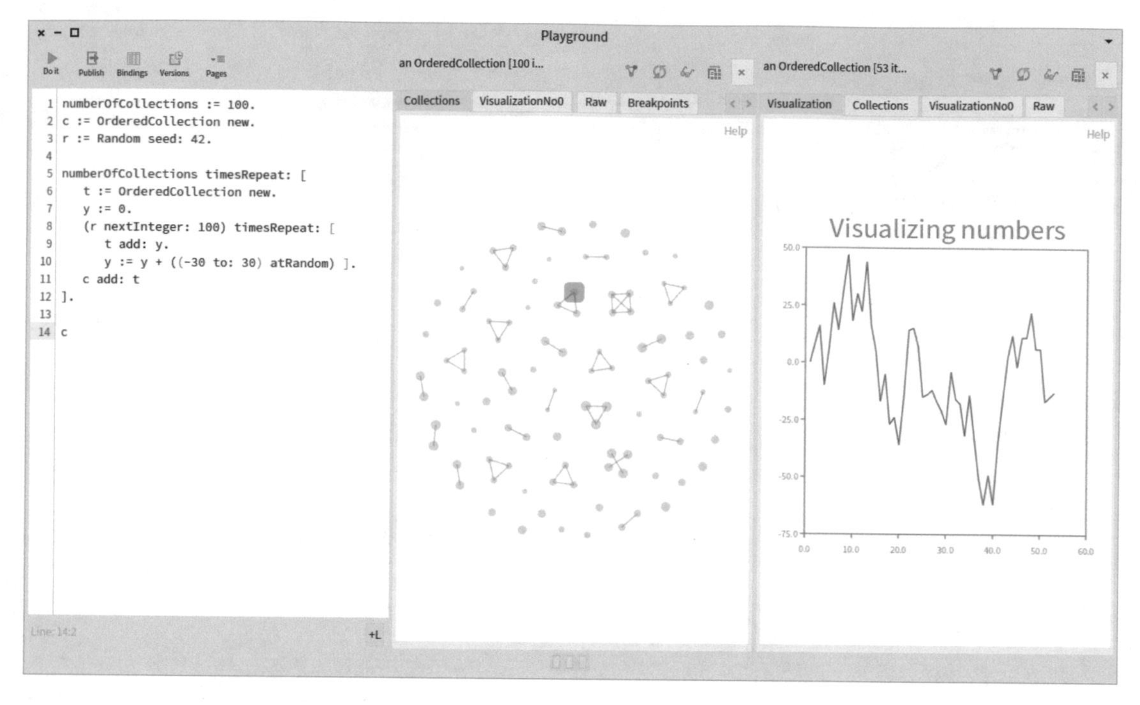

Figure 13-11. Visualizing many nested collections

Figure 13-11 visualizes a larger collection, made of 100 nested collections. Clusters can be easily distinguished in the `Collections` visualization.

What Have You Learned in This Chapter?

Roassal greatly benefits from the programming environment offered by Pharo. The Inspector, easily accessible from the Playground, is a central tool to support incremental and live programming. The chapter explained the following:

- How to add visualizations accessible from the Inspector.
- How to remove the evaluation pane in the Inspector for a particular visualization.
- Examples of the Chart and Mondrian Roassal components.

The chapter defined a few simple visualizations on the `SequenceableCollection` class; however, these very same techniques can be applied to any class from any particular domain.

CHAPTER 14

Reinforcement Learning

Reinforcement learning is an area of machine learning that searches for the behavior of an agent when performing actions in a particular environment. The objective of the behavior is to maximize the notion of cumulative reward by identifying the best action to take for a given environment.

Machine learning is an application of artificial intelligence that provides a system the ability to automatically learn and improve from experience without being explicitly programmed. Reinforcement learning (RL) is an area of machine learning that has made significant contributions in various areas, including robotics and video games. RL is about taking a suitable action to maximize reward in a particular situation.

This chapter illustrates the use of visualization to highlight how reinforcement learning operates on a simple example. This chapter provides a generic implementation of reinforcement learning, called *Q-Learning*, and proposes some visualizations to help illustrate the execution of the algorithm. The chapter is self-contained by providing the complete implementation of Q-Learning and a small game as an illustration.

This chapter is relatively long and illustrates the use of a number of Roassal components:

- *Mondrian* is a high-level API to build visual graphs. The `Roassal3-Mondrian` package contains the whole implementation of Mondrian and numerous examples.
- *Chart* is a high-level API to build charts. The `Roassal3-Chart` package contains all the basic infrastructure to build charts.
- Integration in the Inspector to have multiple visualizations of the same object. All the examples in this chapter are meant to be executed in the Playground, using Cmd+G or Alt+G, which opens an Inspector pane. I recommend you read Chapter 13 before reading this chapter.

A. Bergel, *Agile Visualization with Pharo*, https://doi.org/10.1007/978-1-4842-7161-2_14

All the code provided in this chapter is available at `https://github.com/bergel/AgileVisualizationAPressCode/blob/main/03-01-ReinforcementLearning.txt`.

Implementation Overview

This chapter uses reinforcement learning on a simple and intuitive example, which is a small knight trying to find the exit without encountering monsters. The game is very simple, but it illustrates the need for some visualizations to better understand how Q-Learning, the algorithm used in this implementation, operates.

Reinforcement learning computes the cumulative rewards over a sequence of actions. An exploration step is called an *episode*. Q-Learning and the game implementation span over three classes:

- `RLGrid` defines a grid, which is a simple square area that contains the exit and some monsters.
- `RLState` defines a state, consisting of a grid and the position of the knight.
- `RL` implements Q-Learning, the reinforcement learning algorithm.

Defining the Map

The `RLGrid` class represents the map in which the knight can move around. The grid is a simple matrix containing characters. To encode the content of the map, use the following conventions:

- `$e` represents the exit. The knight is looking for the exit to finish the game.
- `$m` represents a monster. The knight has to learn to avoid monsters.
- `$.` represents an empty cell. The knight can freely move on an empty cells.

RLGrid is defined as follows:

```
Object subclass: #RLGrid
    instanceVariableNames: 'content'
    classVariableNames: ''
    package: 'ReinforcementLearning'
```

The class defines a variable called content, which is a two-dimensional array of characters. A grid is initialized using the following:

```
RLGrid>>initialize
    super initialize.
    self setSize: 2
```

By default, a grid is a square of two cells per side. The setSize: method creates the following map:

```
RLGrid>>setSize: anInteger
    "Set the grid as a square of size anInteger, containing the character $."
    content := (1 to: anInteger) collect: [ :notUsed | Array new: anInteger
    withAll: $. ] as: Array.
    self setExitBottomRight.
```

To keep this code relatively short to fit in a chapter, it uses a number of conventions. The character $. represents a space in which the knight can move around freely. The content of the map may be modified using a number of utility methods. The atPosition: method returns a character contained at a particular position:

```
RLGrid>>atPosition: aPoint
    "Return the character located at a given position"
    ^ (content at: aPoint y) at: aPoint x
```

The map can be modified using the atPosition:put: method, defined as follows:

```
RLGrid>>atPosition: aPoint put: aCharacter
    "Set the aCharacter (value of a cell map) at a given position"
    ^ (content at: aPoint y) at: aPoint x put: aCharacter
```

The exit of the map is set using a dedicated method, invoked by `initialize`:

```
RLGrid>>setExitBottomRight
    "Set the exit position at the bottom right of the grid"
    self atPosition: self extent put: $e
```

The extent of the grid is obtained using:

```
RLGrid>>extent
    "Return a point that represents the extent of the grid"
    ^ content first size @ content size
```

Monsters are defined in a map using the `setMonsters:` method. This method takes as an argument the number of monsters to add, and it is defined as follows:

```
RLGrid>>setMonsters: numberOfMonstersToAdd
    | random leftMonsters s pos nbTries |
    random := Random seed: 42.
    leftMonsters := numberOfMonstersToAdd.
    nbTries := 0.
    s := self extent.
    [ leftMonsters > 0 ] whileTrue: [
        pos := (random nextInteger: s x ) @ (random nextInteger: s y).
        (self atPosition: pos) = $.
            ifTrue: [
                nbTries := 0.
                self atPosition: pos put: $m.
                leftMonsters := leftMonsters - 1 ]
            ifFalse: [
                nbTries := nbTries + 1.
                nbTries > 5 ifTrue: [ ^ self ] ]
    ]
```

The method locates a number of monsters on the map. The method uses a number generator with a seed to deterministically generate a map. The method tries to locate monsters at random positions. If there are too many monsters, the method exits after a number of tries.

You now have to consider a technical aspect of the way Q-Learning is implemented. Grid objects need to be structurally compared to determine whether a particular state was already discovered. Consider this expression:

```
(RLGrid new setSize: 2; setMonsters: 3) = (RLGrid new setSize: 2;
setMonsters: 3)
```

At that stage of the implementation, this expression returns `false`, even if the structures of the two grids are identical:

- The two grids have the same size
- The two grids have the same number of monsters
- Those monsters are located at exactly the same locations, thanks to the seed number defined in the `setMonsters:` method you saw earlier

This expression returns `false` as it's the result of comparing the object identity (i.e., the address in memory of the object). Since evaluating `(RLGrid new setSize: 2; setMonsters: 3)` returns a new object, performing the = operation between two different objects will return `false`. However, in this situation, you want the comparison to return `true` since you are comparing two instances of the identical map. As such, the = needs to operate on the structure of the objects, and not on the object identity, as it does by default. You therefore have to redefine the = and `hash` methods (these two methods are highly intertwined).

The equality can be redefined as follows:

```
RLGrid>>= anotherObject
    "Return true if anotherObject is the same map than myself"
    anotherObject class == self class ifFalse: [ ^ false ].
    ^ content = anotherObject content
```

Two objects are equal if their class and contents are the same. The contents of a grid are obtained with this method:

```
RLGrid>>content
    "Return the grid content, as an array of array of characters"
    ^ content
```

Objects can also be compared through their hash values: two objects that are considered equal (i.e., comparing them with = returns `true`) must have the same hash value. Hash values are intensively used in dictionaries. The QTable is an essential component of the Q-Learning algorithm that maps states and actions to a reward. It is useful to store what the algorithm is learning. In this example, you will model the QTable as a dictionary, and as such, a grid needs to have a proper hash function. Luckily, defining such a hash function is simple in this case:

```
RLGrid>>hash
    "The hash of a grid is the hash of its content"
    ^ content hash
```

After having implemented this method, the following expression evaluates to `true`:

```
(RLGrid new setSize: 2; setMonsters: 3) = (RLGrid new setSize: 2;
setMonsters: 3)
```

New states are created during the exploration of the grid. A copy of the grid is associated with each new state, and you need a way to copy a grid. Pharo provides an easy way to copy any object, by simply defining the `postCopy` method:

```
RLGrid>>postCopy
    "A grid must be properly copied"
    super postCopy.
    content := content copy
```

You can visualize a grid to see the result of the reinforcement learning algorithm. Define a simple method that defines a grid of boxes:

```
RLGrid>>visualize
    | canvas shapes |
    canvas := RSCanvas new.
    shapes := RSBox models: (self content flatCollect: #yourself)
    forEach: [ :s :o |
        s size: 20.
        o = $. ifTrue: [ s color: Color veryVeryLightGray ].
        o = $m ifTrue: [ s color: Color lightRed ].
        o = $e ifTrue: [ s color: Color lightYellow ].
         ].
```

```
    canvas addAll: shapes.
    RSGridLayout new gapSize: 0; lineItemsCount: (self extent x); on:
    shapes.
    shapes translateTopLeftTo: 0 @ 0.
    ^ canvas
```

Each cell of the map is represented as a small, squared box. The color of each box indicates what the cell actually represents. A space is represented as a gray box, a monster is a red box, and the exit is a yellow box. You can make the Inspector show the visualization defined on a grid by simply defining the following:

```
RLGrid>>inspectorVisualization
    <inspectorPresentationOrder: 90 title: 'Visualization'>
    | canvas |
    canvas := self visualize.
    canvas @ RSCanvasController.
    ^ SpRoassal3InspectorPresenter new
        canvas: canvas;
        yourself
```

You can now inspect the following expression by pressing Cmd+G or Alt+G in the Playground (see Figure 14-1):

```
RLGrid new setSize: 5; setMonsters: 5
```

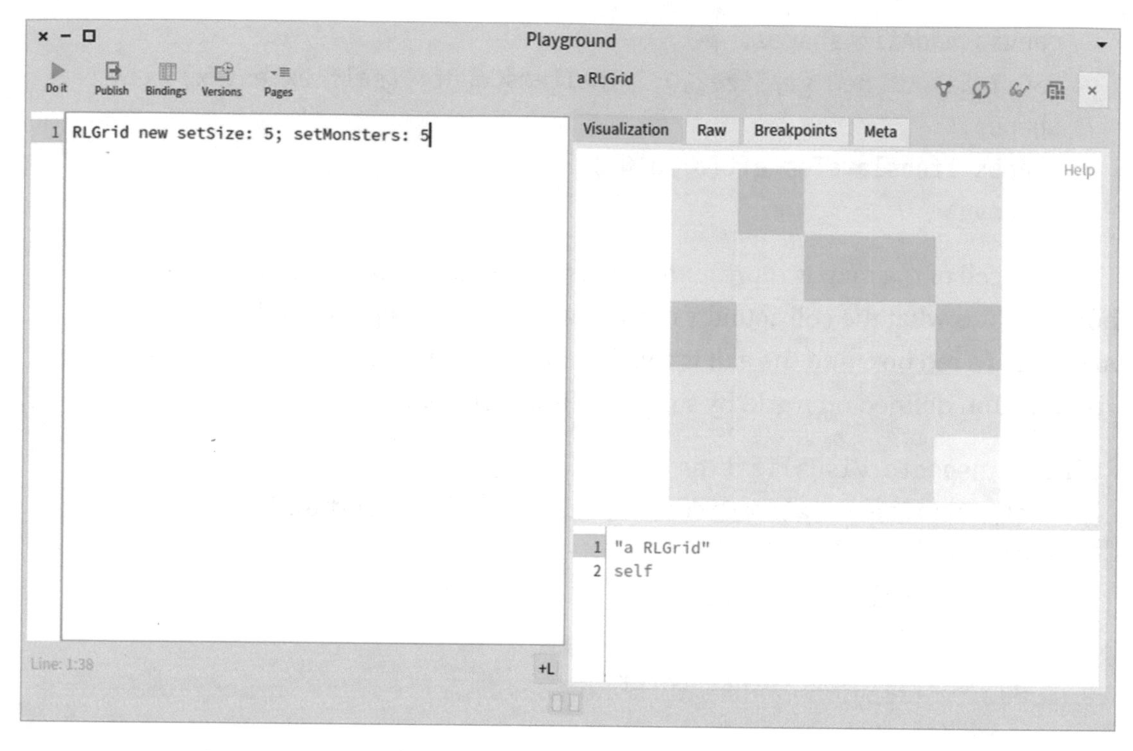

Figure 14-1. *Inspecting a grid*

Modeling State

Modeling the state is the second main component of this chapter. A *state* is a combination of a grid with a position of the knight. The reinforcement learning algorithm operates by discovering states; the actions taken by the knight are expressed using state transitions. The `RLState` class is defined as follows:

```
Object subclass: #RLState
    instanceVariableNames: 'grid position'
    classVariableNames: ''
    package: 'ReinforcementLearning'
```

The default position of the knight is set in the `initialize` method:

```
RLState>>initialize
    super initialize.
    position := 1 @ 1
```

A grid can be set to a state using the following:

```
RLState>>grid: aGrid
    "Set the grid associated to the state"
    grid := aGrid
```

A grid can be obtained from a state using the following:

```
RLState>>grid
    "Return the grid associated to the state"
    ^ grid
```

The position of the knight can be set using the following:

```
RLState>>position: aPoint
    "Set the knight position"
    position := aPoint
```

The knight's position is fetched using the following:

```
RLState>>position
    "Return the knight position"
    ^ position
```

Similar to `RLGrid`, a state needs to be compared with other states. Define the following method:

```
RLState>>= anotherState
    "Two states are identical if (i) they have the same class, (ii) the
    same grid, and (iii) the same position of the knight"
    anotherState class == self class ifFalse: [ ^ false ].
    ^ grid = anotherState grid and: [ position = anotherState position ]
```

The hash value of a state can be obtained using the following:

```
RLState>>hash
    "The hash of a state is a combination of the hash of the grid with the
    hash of the position"
    ^ grid hash bitXor: position hash
```

Obtaining a string representation of a state is handy, especially to represent the QTable:

```
RLState>>printOn: str
    "Give a string representation of a state"
    str nextPutAll: 'S<'.
    str nextPutAll: self hash asString.
    str nextPutAll: '>'.
```

The `printOn:` method produces a string that can be used as a textual label for the state. For example, the `RLState new printString` expression returns a string called `'S<376548926>'`. The numerical value depends on the internal state of the Pharo virtual machine, so you will likely have a different value.

You can now focus on the visualization of the state. The `visualize` method simply renders the grid and the position of the knight, and it is defined as follows:

```
RLState>>visualize
    "Visualize the grid and the knight position"
    | c knightShape |
    c := grid visualize.
    knightShape := RSCircle new size: 15; color: Color blue lighter lighter.
    c add: knightShape.
    knightShape translateTo: self position - (0.5 @ 0.5) * 20.
    ^ c
```

The visualization can now be hooked into the Inspector's framework:

```
RLState>>inspectorVisualization
    <inspectorPresentationOrder: 90 title: 'Visualization'>
    | canvas |
    canvas := self visualize.
    canvas @ RSCanvasController.
    ^ SpRoassal3InspectorPresenter new
        canvas: canvas;
        yourself
```

You can now try the visualization by using Cmd+G/Alt+G on the following script (see Figure 14-2):

```
RLState new
    grid: (RLGrid new setSize: 5; setMonsters: 5)
```

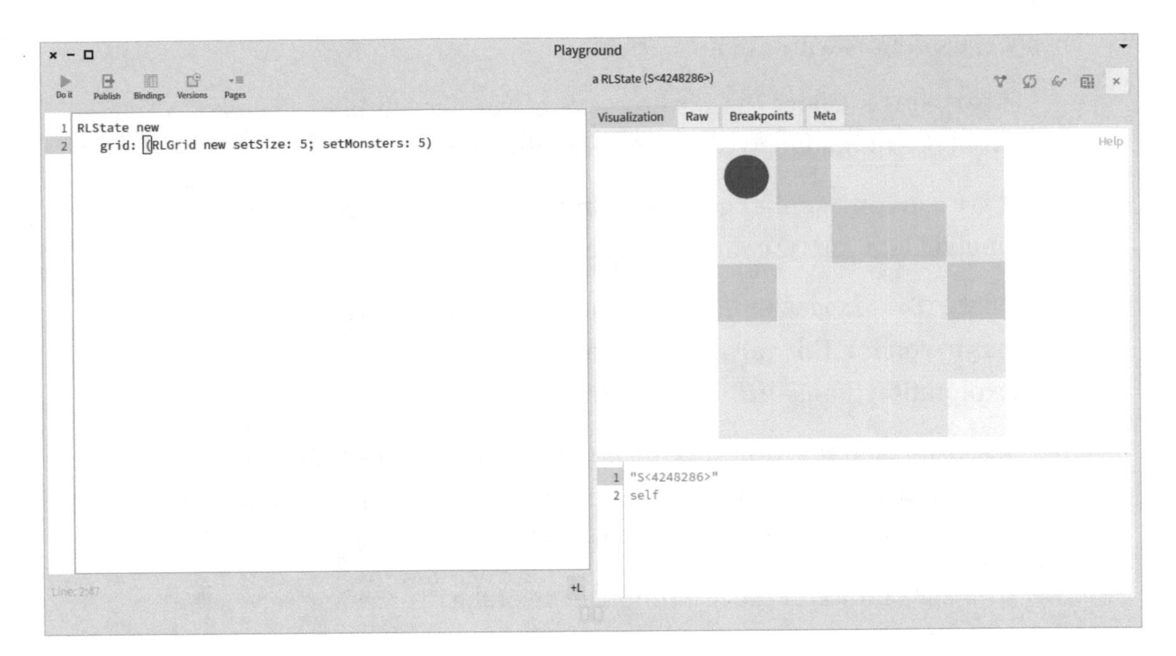

Figure 14-2. *Visualizing a state*

The squared map is five cells wide, the monsters are in light red, and the exit is in yellow.

The Reinforcement Learning Algorithm

The third component of this chapter is the reinforcement algorithm engine itself. The RL class is large and contains many algorithmic parts for which the chapter provides the code without expanding on its rationale. The goal of this chapter is to illustrate how visualizations are beneficial when using reinforcement learning. As such, this chapter is not a tutorial about reinforcement learning. The good news is that there are not many technical parts. First define the RL class as follows:

```
Object subclass: #RL
    instanceVariableNames: 'startState r numberOfEpisodes maxEpisodeSteps
    minAlpha gamma epsilon qTable rewards path stateConnections'
    classVariableNames: ''
    package: 'ReinforcementLearning'
```

The RL class defines a number of variables:

- startState defines the starting state of the algorithm. Each episode (unit of exploration in RL) starts from this state.
- r is a random number generator, initialized with a fixed seed to enable determinism and reproducibility.
- numberOfEpisodes indicates the number of episodes the algorithm has to run for. This value can be considered the amount of exploration the algorithm needs to conduct.
- maxEpisodeSteps indicates the number of steps the knight needs to make during each episode. You need to make sure that this number is large enough to allow the knight to find the exit.
- minAlpha indicates the minimum value of the decreasing learning ramp.
- gamma represents the discount of the reward.
- epsilon represents the probability of choosing a random action when acting.
- qTable is a dictionary that models the QTable and contains states as keys and rewards as values (each actions is associated with a reward).
- rewards lists the accumulated rewards obtained along the episodes. This list indicates how "well" the algorithm is learning.
- path represents the path of the knight to exit the grid.
- stateConnections describes all the transitions between the states.

As mentioned, this class has a mathematical complexity that is not included here. Only a brief description is provided, leaving the job of detailing how Q-Learning operates for another book. The initialization of an RL object is performed using the following:

```
RL>>initialize
    super initialize.
    r := Random seed: 42.
    numberOfEpisodes := 20.
    maxEpisodeSteps := 100.
```

```
    minAlpha := 0.02.
    gamma := 1.0.
    epsilon := 0.2.
    qTable := Dictionary new.
    rewards := OrderedCollection new.
    path := OrderedCollection new.
    stateConnections := OrderedCollection new.
```

The number of episodes can be set using the following:

```
RL>>numberOfEpisodes: aNumber
    "Set the number of exploration we need to perform"
    numberOfEpisodes := aNumber
```

The propensity for exploring the world versus taking the best action is set by the `epsilon` value:

```
RL>>epsilon: aFloat
    "Set the probability to explore the world. The argument is between 0.0
    and 1.0. A value close to 0.0 favors choosing an action that we know is
    a good one (thus reducing the exploration of the grid). A value close
    to 1.0 favors the world exploration instead."
    epsilon := aFloat
```

Set the maximum number of steps for each episode as follows:

```
RL>>maxEpisodeSteps: anInteger
    "Indicate how long an exploration can be"
    maxEpisodeSteps := anInteger
```

The value of `maxEpisodeSteps` has a default value of `100`, as mentioned in `initialize`. This value is high enough for maps having fewer than 100 cells (i.e., a grid smaller than 10 x 10).

For a given state and action, you must determine the behavior of the knight, compute the reward gained by performing the action, and indicate whether the game is over. Define the `act:action:` method for that purpose:

```
RL>>act: aState action: action
    "Produce a new tuple {stable . reward . isDone}"
    | reward newGrid p gridItem isDone newState |
```

```
    p := self moveKnight: aState action: action.
    gridItem := aState grid atPosition: p.
    newGrid := aState grid copy.
    gridItem = $m ifTrue: [ reward := -100. isDone := true ].
    gridItem = $e ifTrue: [ reward := 1000. isDone := true ].
    ('me' includes: gridItem)
        ifFalse: [ reward := -1. isDone := false ].
    newState := RLState new grid: newGrid; position: p.
    stateConnections add: aState -> newState.
    ^ { newState . reward . isDone }
```

The knight can move by one cell in any of the four directions. A move is represented as an action, itself described as a numerical value ranging from 1 to 4. Define the actions method, which simply returns the list of the actions:

```
RL>>actions
    "Return the considered actions"
    ^ #(1 2 3 4)
```

You represent an action as a number since it is convenient to access the Q values from the QTable. For a given state, the knight needs to take an action. Either this action is random or it is based on the exploration made by the knight. The chooseAction: method is designed for that purpose:

```
RL>>chooseAction: state
    "Choose an action for a given state"
    ^ r next < epsilon
        ifTrue: [ self actions atRandom: r ]
        ifFalse: [
            "Return the best action"
            (self qState: state) argmax ]
```

Once an action has been determined, the knight needs to move accordingly. It is important for the knight to not exit the grid:

```
RL>>moveKnight: state action: action
    "Return the new position of a car, as a point. The action is a number
    from 1 to 4.
    return a new position"
```

```
    | delta |
    delta := { 0@ -1 . 0@1 . -1@0 . 1@0 }
                at: action ifAbsent: [ self error: 'Unknown action' ].
    ^ ((state position + delta) min: state grid extent) max: 1 @ 1
```

Once the knight has explored enough, you can use what it has learned to find the exit. The play method determines a list of actions to be taken to reach the exit:

```
RL>>play
    "Return the position of the car"
    | currentState isDone actions tuple maxNumberOfSteps numberOfSteps |
    currentState := startState.
    isDone := false.
    path := OrderedCollection new.
    path add: currentState position.
    maxNumberOfSteps := 100.
    numberOfSteps := 0.
    [ isDone not and: [ numberOfSteps < maxNumberOfSteps ] ] whileTrue: [
        actions := self qState: currentState.
        tuple := self act: currentState action: actions argmax.
        currentState := tuple first.
        path add: currentState position.
        isDone := tuple third.
        numberOfSteps := numberOfSteps + 1.
    ].

    ^ path asArray
```

For a given state, the rewards associated with the actions are useful to take an action:

```
RL>>qState: state
    "Return the rewards associated to a state. If the state is not in the
    qTable, we create it"
    qTable at: state ifAbsentPut: [ Array new: self actions size withAll: 0 ].
    ^ qTable at: state
```

A particular reward can be accessed from a state and an action:

```
RL>>qState: state action: action
    "For a particular state, return the reward of an action. If the state
    is not in the qTable, we create it"
    qTable at: state ifAbsentPut: [ (1 to: self actions size) collect:
    [ :nU | 0 ] ].
    ^ (qTable at: state) at: action
```

The core of the algorithm is defined by the run method:

```
RL>>run
    "This method is the core of the Q-Learning algorithm"
    | alphas currentState totalReward alpha isDone currentAction tuple
      nextState currentReward |
    alphas := (minAlpha to: 1.0 count: numberOfEpisodes) reversed.
    rewards := OrderedCollection new.
    1 to: numberOfEpisodes do: [ :e |
        currentState := startState.
        totalReward := 0.
        alpha := alphas at: e.
        isDone := false.
        maxEpisodeSteps timesRepeat: [
            isDone ifFalse: [
                currentAction := self chooseAction: currentState.
                tuple := self act: currentState action: currentAction.
                nextState := tuple first.
                currentReward := tuple second.
                isDone := tuple third.
                totalReward := totalReward + currentReward.

                "The Bellman equation"
                (self qState: currentState) at: currentAction put: (
                    (self qState: currentState action: currentAction)
                    + (alpha * (currentReward + (gamma * (self qState:
                    nextState) max) - (self qState: currentState action:
                    currentAction)))).
```

```
                currentState := nextState
            ]
        ].
        rewards add: totalReward.
    ].
    rewards := rewards asArray.
    ^ rewards
```

This method implements the Bellman equation, which is the core of the algorithm. The algorithm decays the learning rate alpha at every episode. As the knight explores more of the environment, it will believe that there is not much to learn.

To simplify the examples you will see shortly, you can define the `setInitialGrid:` method to set the grid in the starting state:

```
RL>>setInitialGrid: aGrid
    "Set the grid used in the initial state"
    startState := RLState new grid: aGrid
```

The core of reinforcement learning algorithm is now implemented. You just need to define methods to visualize some aspects of the algorithm's executions. First the QTable is an important aspect of the algorithm to visualize. Define the `visualizeQTable` method for that purpose:

```
RL>>visualizeQTable
    | c state values allBoxes sortedAssociations |
    c := RSCanvas new.

    c add: (RSLabel text: 'State').
    c add: (RSLabel text: '^').
    c add: (RSLabel text: 'V').
    c add: (RSLabel text: '<').
    c add: (RSLabel text: '>').

    sortedAssociations := qTable associations reverseSortedAs: [ :assoc |
    assoc value average ].
    sortedAssociations do: [ :assoc |
        state := RSLabel model: assoc key.
        values := RSBox
```

```
                                models: (assoc value collect: [ :v | v round: 2 ])
                                forEach: [ :s :m | s extent: 40 @ 20 ].
        c add: state.
        c addAll: values.
    ].
    RSCellLayout new lineItemsCount: 5; gapSize: 1; on: c shapes.
    allBoxes := c shapes select: [ :s | s class == RSBox ].
    RSNormalizer color
        shapes: allBoxes;
        from: Color red darker darker; to: Color green darker darker;
        normalize.
    allBoxes @ RSLabeled middle.
    ^ c @ RSCanvasController
```

The method represents the QTable as a table in which columns are the rewards per actions, and each row is a state. You use a cell layout to locate all the values in the grid. The cell layout arranges all the shapes to be lined up, both vertically and horizontally. This visualization is hooked into the Inspector using the following:

```
RL>>inspectorQTable
    <inspectorPresentationOrder: 90 title: 'QTable'>

    ^ SpRoassal3InspectorPresenter new
        canvas: self visualizeQTable;
        yourself
```

The evaluation panel is not relevant in this case and can be removed using the following:

```
RL>>inspectorQTableContext: aContext
    aContext withoutEvaluator
```

The evolution of the rewards along the episodes can be represented using a chart. The `visualizeReward` method uses the Chart component of Roassal:

```
RL>>visualizeReward
    | c plot |
    c := RSChart new.
    plot := RSLinePlot new.
    plot y: rewards.
```

```
    c addPlot: plot.
    c addDecoration: (RSChartTitleDecoration new title: 'Reward evolution';
    fontSize: 20).
    c xlabel: 'Episode' offset: 0 @ 10.
    c ylabel: 'Reward' offset: -20 @ 0.
    c addDecoration: (RSHorizontalTick new).
    c addDecoration: (RSVerticalTick new).
    c build.
    ^ c canvas
```

The chart to visualize the reward is composed of a line plot. You plot all the values contained in the `rewards` variable. A number of decorations are added to improve the appearance of the chart. The two axes have a label and some ticks. The reward visualization is integrated in the Inspector with the following:

```
RL>>inspectorReward
    <inspectorPresentationOrder: 90 title: 'Reward'>

    ^ SpRoassal3InspectorPresenter new
        canvas: self visualizeReward;
        yourself
```

Since there is no need for an evaluation pane, you can disable it:

```
RL>>inspectorRewardContext: aContext
    aContext withoutEvaluator
```

The path of the knight can be visualized using the following method:

```
RL>>visualizeResult
    "Assume that the method play was previously invoked"
    | c s |
    self play.
    c := startState visualize.
    path do: [ :p |
        s := RSCircle new size: 5; color: Color blue lighter lighter.
        c add: s.
        s translateTo: p - (0.5 @ 0.5) * 20.
    ].
    ^ c @ RSCanvasController
```

The integration in the Inspector is done using the following:

```
RL>>inspectorResult
    <inspectorPresentationOrder: 90 title: 'Result'>

    ^ SpRoassal3InspectorPresenter new
        canvas: self visualizeResult;
        yourself
```

Since there is no need to have an evaluation pane when looking at the result, you can disable it using the following:

```
RL>>inspectorResultContext: aContext
    aContext withoutEvaluator
```

Generally speaking, a graph is a set of nodes connected by edges. In this context, the reinforcement learning algorithm builds a graph made of states and actions connecting those states. Exploring the map is expressed by adding new states to the graph, and identifying an ideal path through the map is the result of determining adequate rewards for each action and each state. Visualizing the graph built by the algorithm is relevant to understanding how the exploration went during the learning phase. You therefore need to build the `visualizeGraph` method:

```
RL>>visualizeGraph
    | s allStates d m |
    s := stateConnections asSet asArray.
    d := Dictionary new.
    s do: [ :assoc |
        (d at: assoc key ifAbsentPut: [ OrderedCollection new ]) add:
        assoc value ].

    allStates := qTable keys.
    m := RSMondrian new.
    m shape circle.
    m nodes: allStates.
    m line connectToAll: [ :aState | d at: aState ifAbsent: [ #() ] ].
    m layout force.
    m build.
    ^ m canvas.
```

The visualizeGraph method uses the RSMondrian class, which is a very useful class offered by Roassal to build graphs. A circle shape is selected using m shape circle. During the learning phase, the stateConnections variable is filled with all the associations between states and the qTable variable is filled with the rewards associated with each action. The qTable key expression returns all the states that were created during the learning phase. For a given state aState, the expression d at: aState ifAbsent: [#()] returns the list of connected states (i.e., a directed going from aState to the connected states).

The graph can be accessible from the Inspector using the following method:

```
RL>>inspectorGraph
    <inspectorPresentationOrder: 90 title: 'State graph'>

    ^ SpRoassal3InspectorPresenter new
        canvas: self visualizeGraph;
        yourself
```

Since an evaluator context does not make much sense to have for visualizeGraph, you can remove it using the method:

```
RL>>inspectorGraphContext: aContext
    aContext withoutEvaluator
```

You can define a complementary visualization to associate some metrics with each state. Consider the following script:

```
RL>>visualizeWeightedGraph
    | s allStates d m |
    s := stateConnections asSet asArray.
    d := Dictionary new.
    s do: [ :assoc |
        (d at: assoc key ifAbsentPut: [ OrderedCollection new ]) add:
        assoc value ].

    allStates := qTable keys.
    m := RSMondrian new.
    m shape circle.
    m nodes: allStates.
    m line connectToAll: [ :aState | d at: aState ifAbsent: [ #() ] ].
```

```
m normalizeSize: [ :aState | (qTable at: aState) average ] from: 5 to: 30.
m normalizeColor: [ :aState | (qTable at: aState) max ] from: Color
gray to: Color green.
m layout force.
m build.
^ m canvas.
```

The visualization defined in visualizeWeightedGraph is very similar to the one in visualizeGraph. The only difference is that it has some metrics associated with the size and color of each state. The size of a state indicates the average reward on the four actions: the higher the average is, the larger the state. As such, a large state contributes to reaching the exit.

The color of a state indicates the maximum reward among the four actions: if an action with a high reward (e.g., if the knight is next to the exit) can be performed on a state, this state will be colored in green. The weighted graph visualization is hooked into the Inspector using the following:

```
RL>>inspectorWeightedGraph
    <inspectorPresentationOrder: 90 title: 'Weighted state graph'>

    ^ SpRoassal3InspectorPresenter new
        canvas: self visualizeWeightedGraph;
        yourself
```

Since there is no need to have the evaluation pane, you can disable it:

```
RL>>inspectorWeightedGraphContext: aContext
    aContext withoutEvaluator
```

You have now implemented the whole reinforcement learning algorithm and four visualizations:

- *Result visualization* shows the path taken by the knight to escape the grid.
- *QTable* renders the Q table, in which each cell of the matrix contains the reward of an action in a given state.
- *State graph* renders the graph explored by the algorithm (states are nodes and actions are edges).

- *Weighted state graph* highlights the importance of the state to contribute to exiting the grid.

The visualization is accessible when inspecting an instance of the RL class.

Running the Algorithm

Now that you have implemented the reinforcement learning algorithm and have added the necessary script to have the knight exit the grid, it is time to try it out! Evaluate the following code snippet in the Playground by pressing Cmd+G/Alt+G (see Figure 14-3):

```
rl := RL new.
rl setInitialGrid: RLGrid new.
rl run.
rl
```

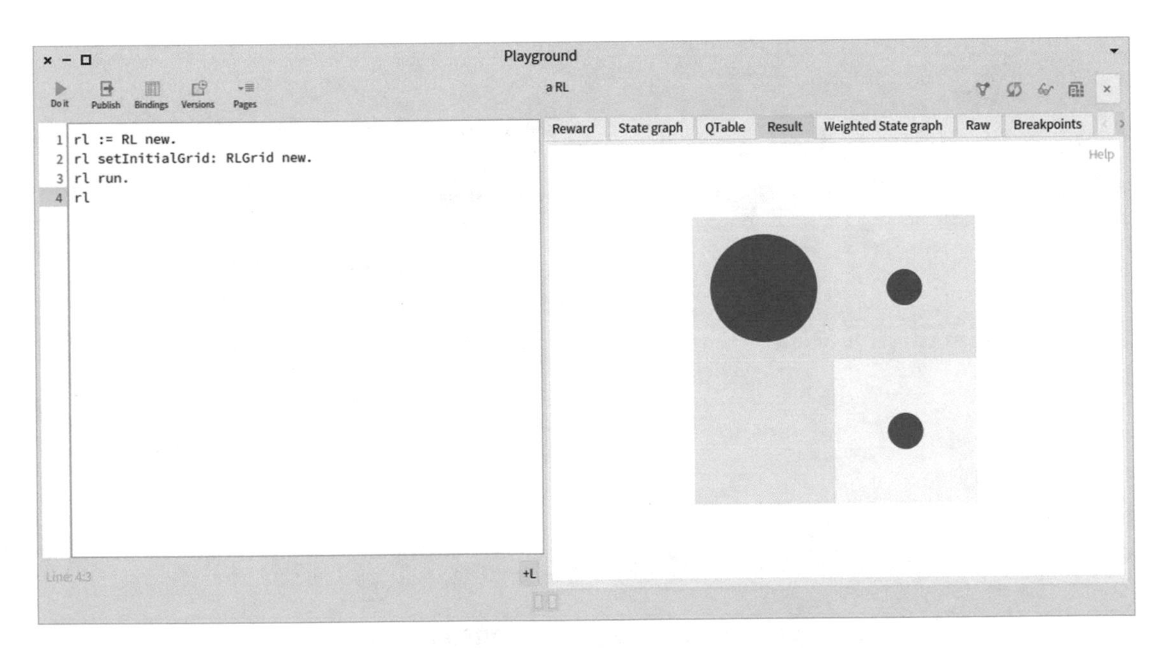

Figure 14-3. *Reaching the exit in the simplest possible grid*

Evaluating the script split the Playground in two. The right side of the Playground inspects the result of executing the script, an RL object in this case. The grid used in the example has a side of two cells, and no monster inhabits the map. The large blue dot indicates the initial position of the knight and the smaller dots indicate the path. In Figure 14-2, you can see that the knight needs to move two cells to reach the yellow cell, which represents the exit.

The example seems trivial, but actually, it is not. If you closely look at the program, you never "told" the knight that it has to exit the map! The knight discovered that exiting the map is the best way to maximize the cumulative reward. By randomly choosing some actions, it discovered the map. Note that the only cell the knight can see is the one that it's standing on.

Now try running the reinforcement learning algorithm on a larger example. Consider the following code snippet (see Figure 14-4):

```
rl := RL new.
rl setInitialGrid: (RLGrid new setSize: 5; setMonsters: 2).
rl run.
rl
```

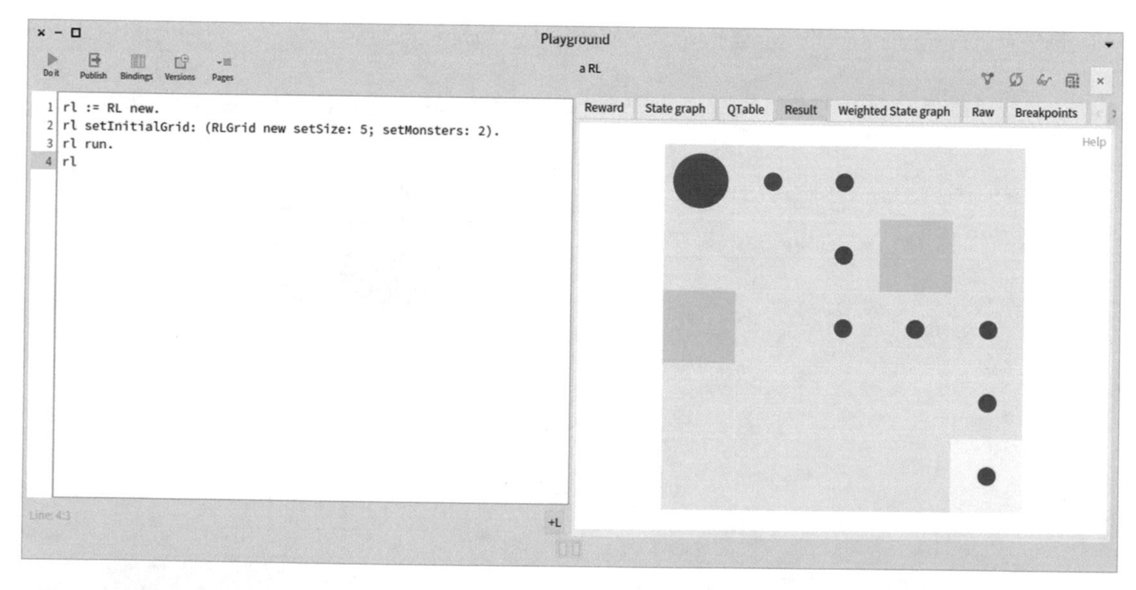

Figure 14-4. *Exiting the grid while avoiding monsters*

Figure 14-4 shows the path taken by the knight to exit the map while avoiding the two monsters. Selecting the Reward tab shows the evolution of the reward along the episodes (see Figure 14-5).

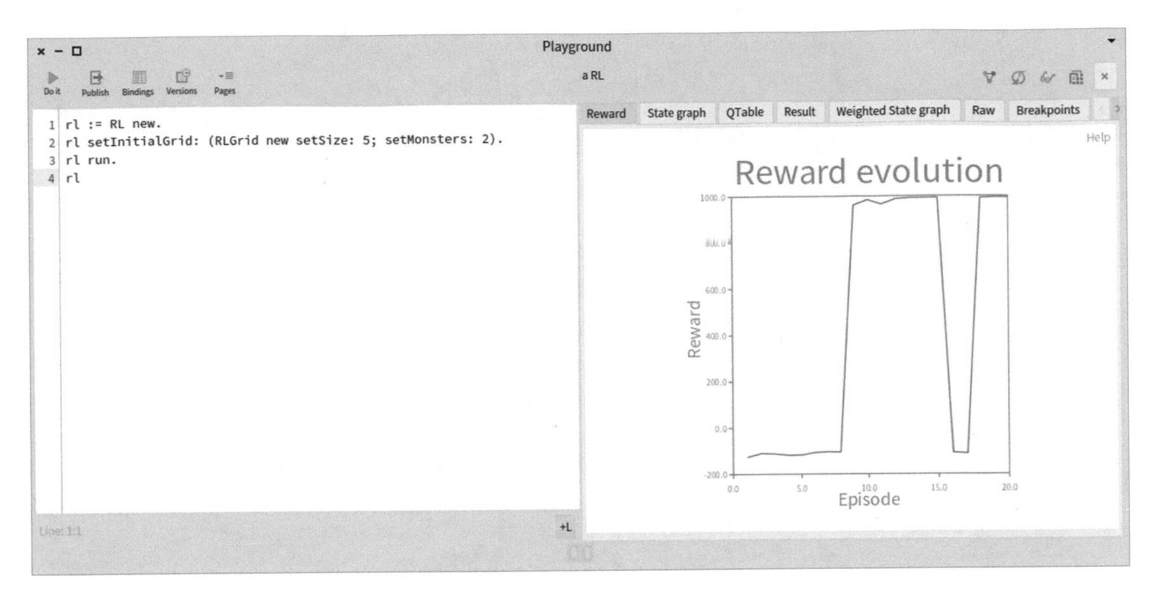

***Figure 14-5.** The reward evolution indicates that the algorithm is learning*

The reinforcement learning algorithm balances the exploration of the map versus taking good actions that were previously learned. Figure 14-5 shows sudden drops in the reward evolution. This indicates an accumulation of explorations that lead the knight to bump into monsters, the result of some random explorations. The `epsilon` value indicates the likelihood to explore the map: a low value (i.e., positive float number close to 0) indicates little exploration, favoring recalling what it has learned. Having a high epsilon (i.e., a float value slightly below 1.0) favors random exploration. You can verify the effect of `epsilon` using the following script (see Figure 14-6):

```
rl := RL new.
rl setInitialGrid: (RLGrid new setSize: 5; setMonsters: 2).
rl epsilon: 0.01.
rl run.
rl
```

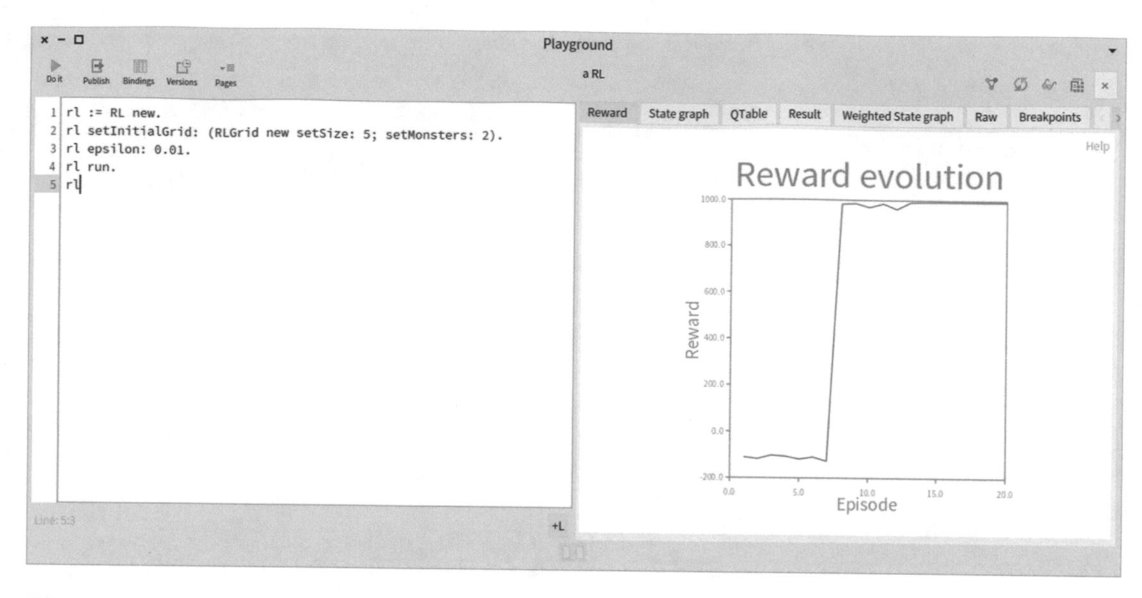

***Figure 14-6.** Reducing the epsilon value prevents sudden drops in the reward evolution*

The default value of epsilon is 0.2, as defined in the `initialize` method. Setting a significantly smaller value, such as 0.01 in the script, reduces the discovery of new path in favor of reusing what the knight has learned. As such, there is no sudden drop in the reward evolution chart. Having the reward at a high level indicates that the Q-Learning has actually learned the best action to take in the relevant states.

The State Graph tab in the Playground represents the graph composed of the explored states (see Figure 14-7).

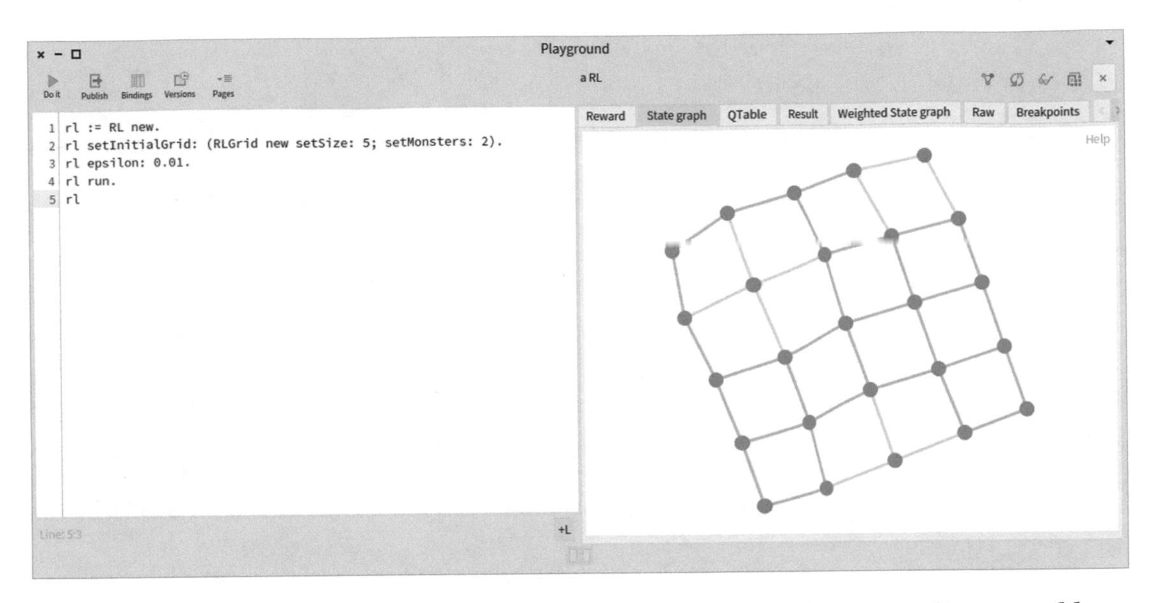

***Figure 14-7.** Graph illustrating the connections between the states discovered by the reinforcement learning algorithm*

A state is a combination of a grid and a position. The graph shown in Figure 14-7 has the shape of a square. This indicates that the knight has visited every position of the map, even the one containing the monsters. If you reduce the amount of iterations used when running the algorithm, fewer states will be discovered. You can verify this using the new script (see Figure 14-8):

```
rl := RL new.
rl setInitialGrid: (RLGrid new setSize: 5; setMonsters: 2).
rl epsilon: 0.01.
rl numberOfEpisodes: 7.
rl run.
rl
```

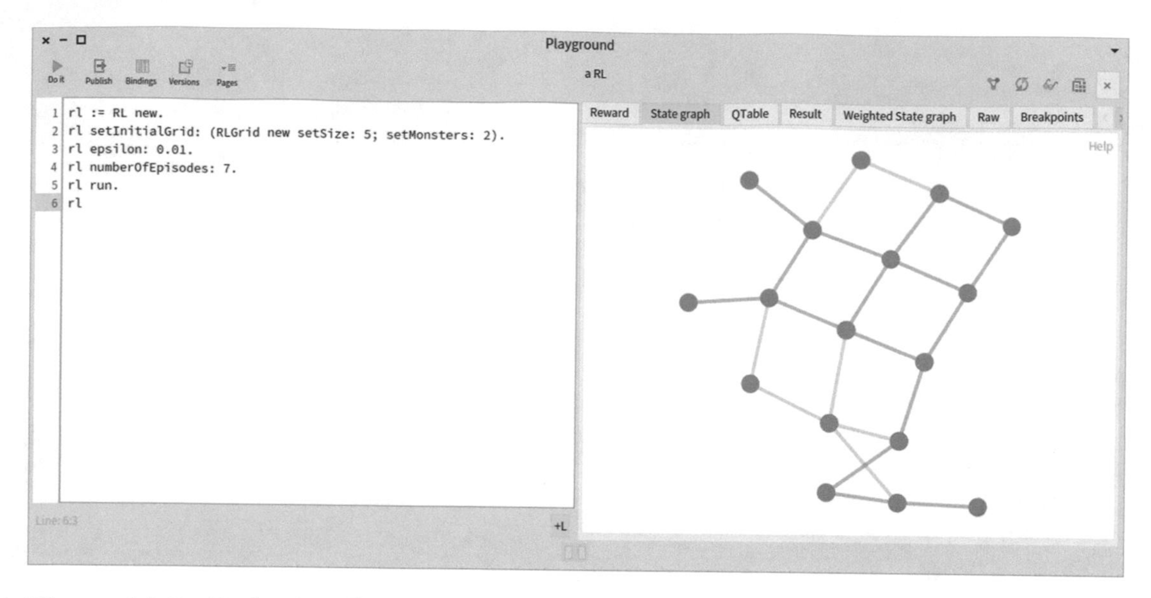

***Figure 14-8.** Reducing the exploration is translated into discovering fewer states*

The number of episodes set by default is 20, as indicated in the `initialize` method. Reducing this value to 7 implies fewer iterations when running the algorithms, and as such, fewer states are discovered. Figure 14-8 indicates this situation by having fewer nodes and fewer edges. Note that the knight cannot find the exit of the map.

The QTable tab shows the matrix of the reward values for a state and an action (see Figure 14-9).

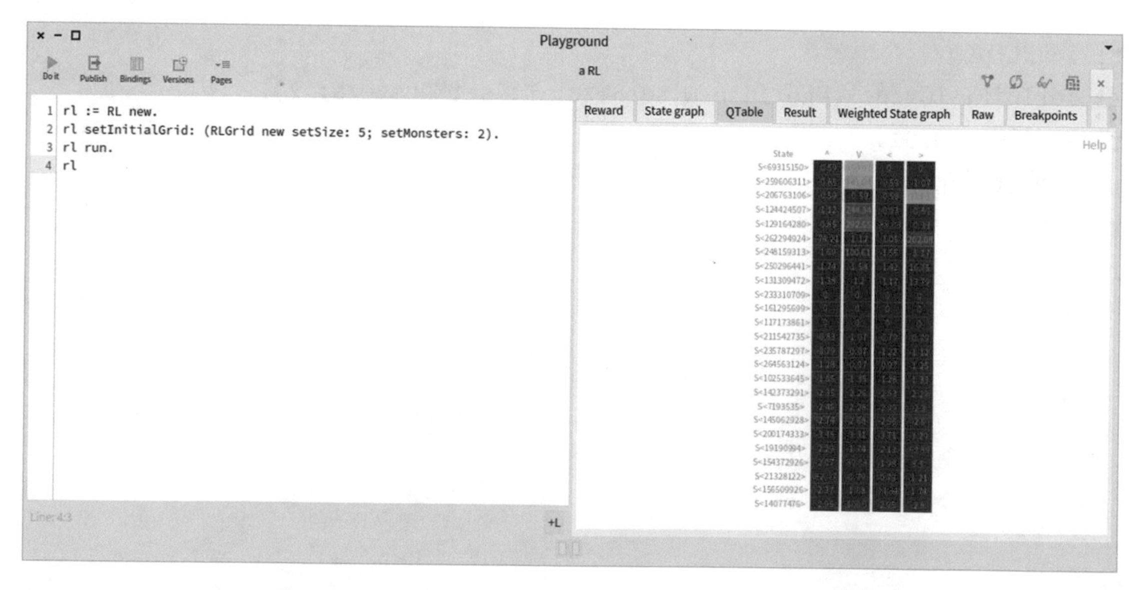

***Figure 14-9.** The QTable matrix*

The red indicates the actions that are not beneficial. A state can be inspected by clicking it. The first state has a green instance action, representing the action to go down. Clicking it reveals that the state corresponds to when the knight is just above the exit. Going down one cell in that case leads to a very high reward (see Figure 14-10).

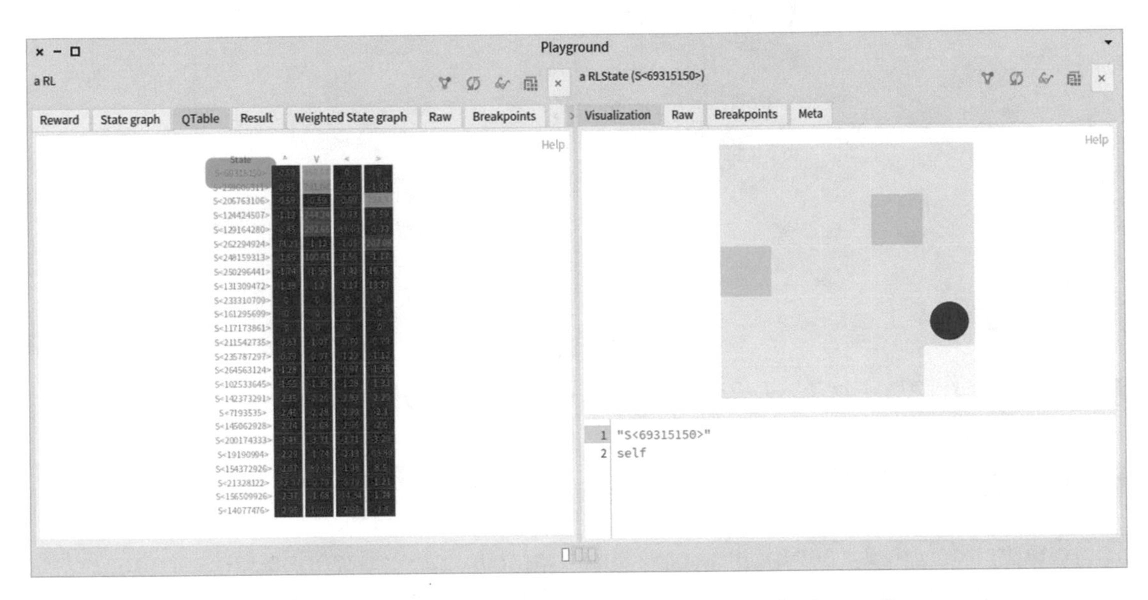

Figure 14-10. *States in the QTable can be inspected by clicking them*

The weighted state graph illustrates the relevance of a state to contribute to reaching the exit. Consider the script given previously (see Figure 14-11):

```
rl := RL new.
rl setInitialGrid: (RLGrid new setSize: 5; setMonsters: 2).
rl run.
rl
```

Large green circles represent states that contribute to maximizing the cumulative reward and therefore reaching the exit.

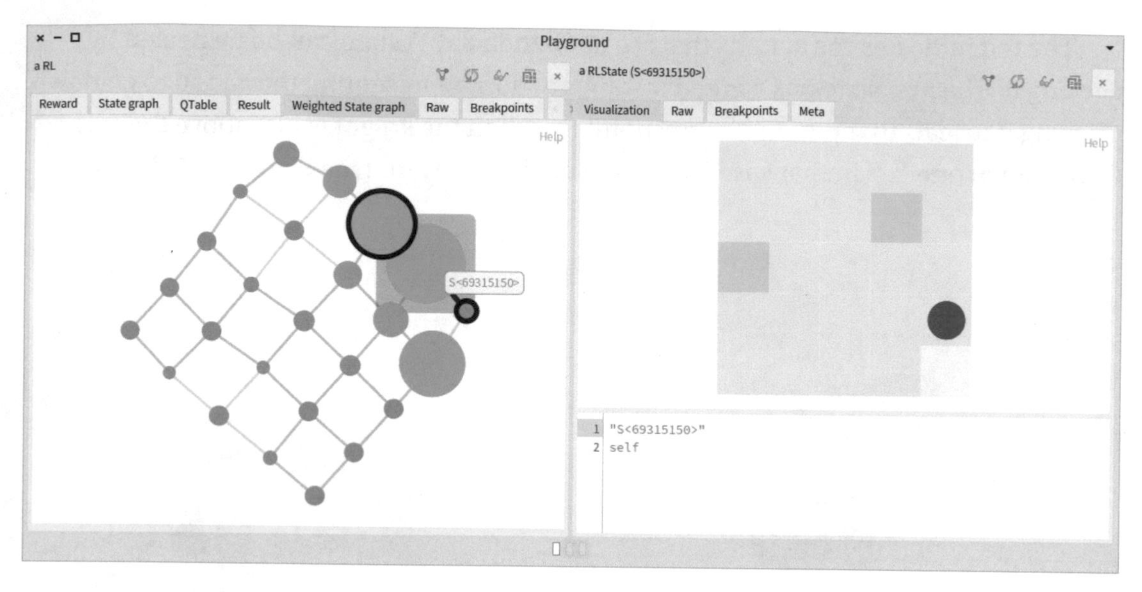

Figure 14-11. *Normalized metrics applied to the graph of states*

What Have You Learned in This Chapter?

This chapter detailed a compelling case study for using visualizations. Q-Learning is a very popular algorithm that belongs to the family of reinforcement learning. Q-Learning, as with most machine learning algorithms, is best understood with visual representations. The visualizations were realized using dedicated and expressive components. This chapter included:

- An illustration of using Mondrian to build and render graphs.
- Shapes in the Mondrian graph normalized using some metrics.
- Chart built using the API provided by Roassal.
- A complete implementation of the Q-Learning algorithm with a classic scenario.

CHAPTER 15

Generating Visualizations From GitHub

GitHub is the most popular source code repository used at the time this chapter was written. A repository on GitHub contains historical information about a project's source code and it lets anyone contribute through commits. GitHub Actions is a service proposed by GitHub that helps you automate tasks in a software development lifecycle and therefore support continuous delivery. GitHub Actions associates an action to a particular event, which is typically executing a script at each commit or pull request. The GitHub Actions mechanism opens doors to many advanced applications, in particular producing visualizations along a software evolution.

After a brief introduction to GitHub Actions, the chapter gives examples of using GitHub Actions to execute unit tests and to generate software visualizations. This chapter involves very little programming since it mostly sets up configuration and workflows, and defines scripts through dedicated programming languages. However, it highlights how Pharo and Roassal can be a significant asset to continuous delivery, as supported by GitHub. You are expected to be familiar with Git and GitHub.

I strongly encourage you to copy and paste the scripts from the accompanying code files. Many scripts given in this chapter are written in YAML, and it is unfortunately too easy to make mistakes when writing scripts. All the code and scripts provided in this chapter are available at `https://github.com/bergel/AgileVisualizationAPressCode/blob/main/03-02-ContinuousIntegration.txt`.

A. Bergel, *Agile Visualization with Pharo*, https://doi.org/10.1007/978-1-4842-7161-2_15

Requirements

To replicate what is presented in this chapter, you should have a GitHub repository with a Pharo application. The application must have a baseline or a way to be installed via a code snippet. The application has to contain unit tests since you will see, as a first example of GitHub Actions, a way to run unit tests at each commit. The application can be small or large, it does not matter.

This chapter uses the `https://github.com/bergel/ReinforcementLearning` repository to illustrate various concepts. However, all the scripts presented in this chapter are meant to be replicated on any arbitrary Pharo project kept on GitHub.

Creating a Workflow

The workflow is an automated procedure that can be added to a GitHub repository. A workflow is made up of a number of jobs and is triggered by an action, typically pushing on a repository. A workflow can be used to build, test, and produce reports.

You can create a workflow from the online GitHub interface of the project by clicking the Actions tab (see Figure 15-1).

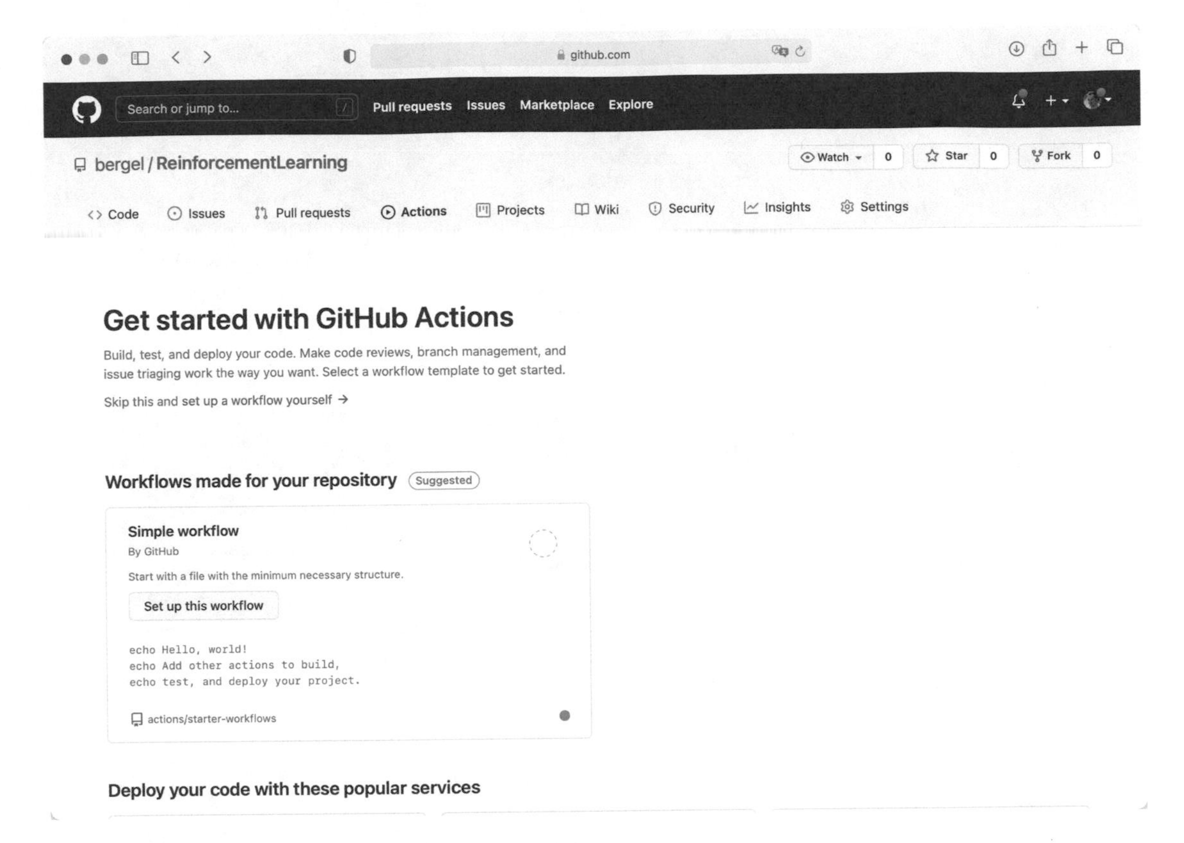

Figure 15-1. *The Actions tab on GitHub*

This web page offers many workflow templates to address needs such as deploying an application to the Microsoft Azure cloud or Amazon AWS. In this case, you will set up a small workflow that simply executes a Pharo script. Click the Setup This Workflow button to enter a workflow definition (see Figure 15-2).

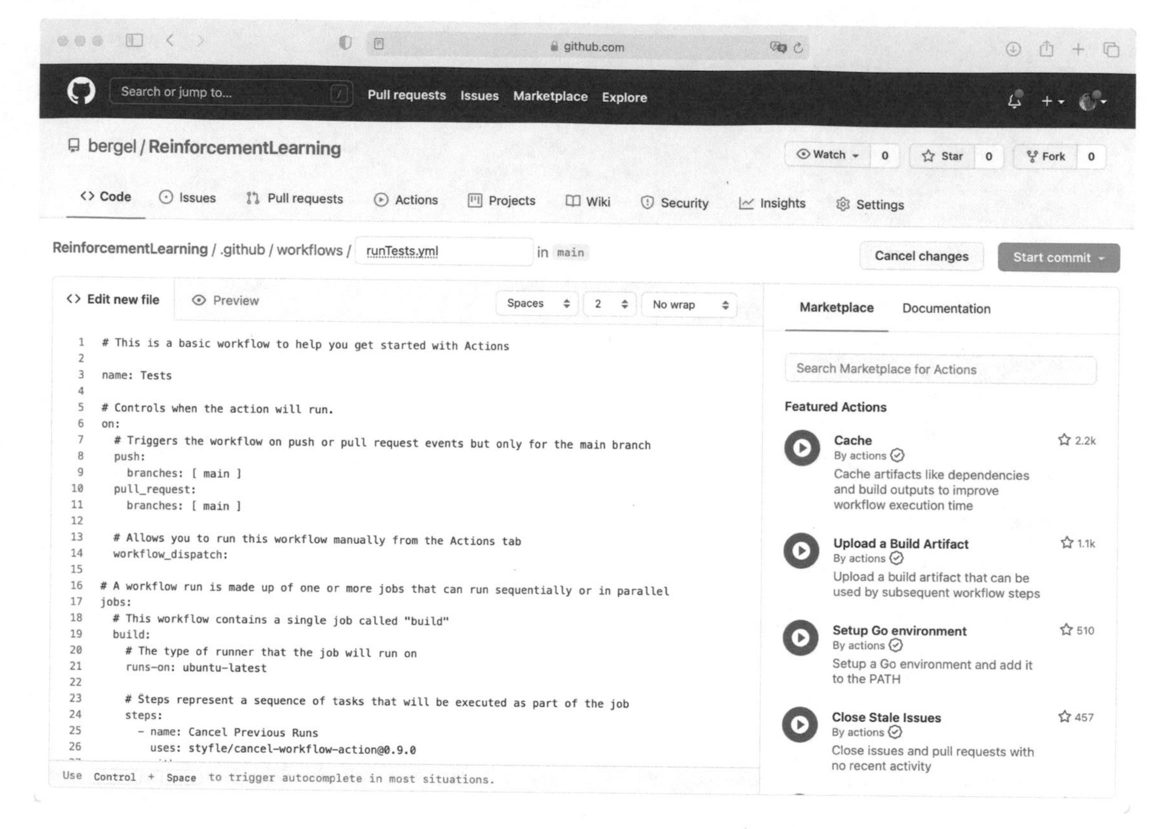

***Figure 15-2.** Editing the workflow*

You can name the workflow by filling in the text field. In this example, it's called `runTests.yml`, as it describes a workflow that will run unit tests. On the top right, the workflow can be committed by pressing the green Start Commit button and the Commit New File button. You can leave the default implementation for now. This workflow does not do much with test executions, but you will modify it shortly.

After having committed the file, a new file called `.github/workflows/runTests.yml` is part of the repository. The file follows the YAML format, which is a file format comparable to XML and JSON, but designed to be human-readable. YAML is often used to write configuration files.

The `runTests.yml` file, as provided by the template, is defined as follows:

```
# This is a basic workflow to help you get started with Actions

name: CI

# Controls when the workflow will run
```

```
on:
  # Triggers the workflow on push or pull request events but only for the
    main branch
  push:
    branches: [ main ]
  pull_request:
    branches: [ main ]

  # Allows you to run this workflow manually from the Actions tab
  workflow_dispatch:

# A workflow run is made up of one or more jobs that can run sequentially
or in parallel
jobs:
  # This workflow contains a single job called "build"
  build:
    # The type of runner that the job will run on
    runs-on: ubuntu-latest

    # Steps represent a sequence of tasks that will be executed as part of
      the job
    steps:
      # Checks out your repository under $GITHUB_WORKSPACE, so your job can
        access it
      - uses: actions/checkout@v2

      # Runs a single command using the runners shell
      - name: Run a one-line script
        run: echo Hello, world!

      # Runs a set of commands using the runners shell
      - name: Run a multi-line script
        run: |
          echo Add other actions to build,
          echo test, and deploy your project.
```

The file defines a GitHub Actions workflow. The name of the work is `CI`, and it is triggered upon `push` and `pull_request` events. The list of steps given at the end of the file is the relevant part of the workflow definition to focus on for now. The two last steps—called `Run a one-line script` and `Run a multi-line script`—contain some shell script echo commands, which simply print a particular string to the standard output.

You may have a slightly different template if you create a workflow on an existing GitHub repository.

For example, in your project, the main branch may be called `master` instead of `main`. In 2020, GitHub considered the term *master* to be harmful and antiquated and opted to rename the principal branch in a GitHub repository *main*.

Trying the Workflow

At each pull and pull request event, the steps described in the workflow are executed in a virtual computer. If you modify the `README.md` file or simply commit a change in your application, the steps defined in the workflow are executed. Clicking the Actions tab lists the executed workflow (see Figure 15-3).

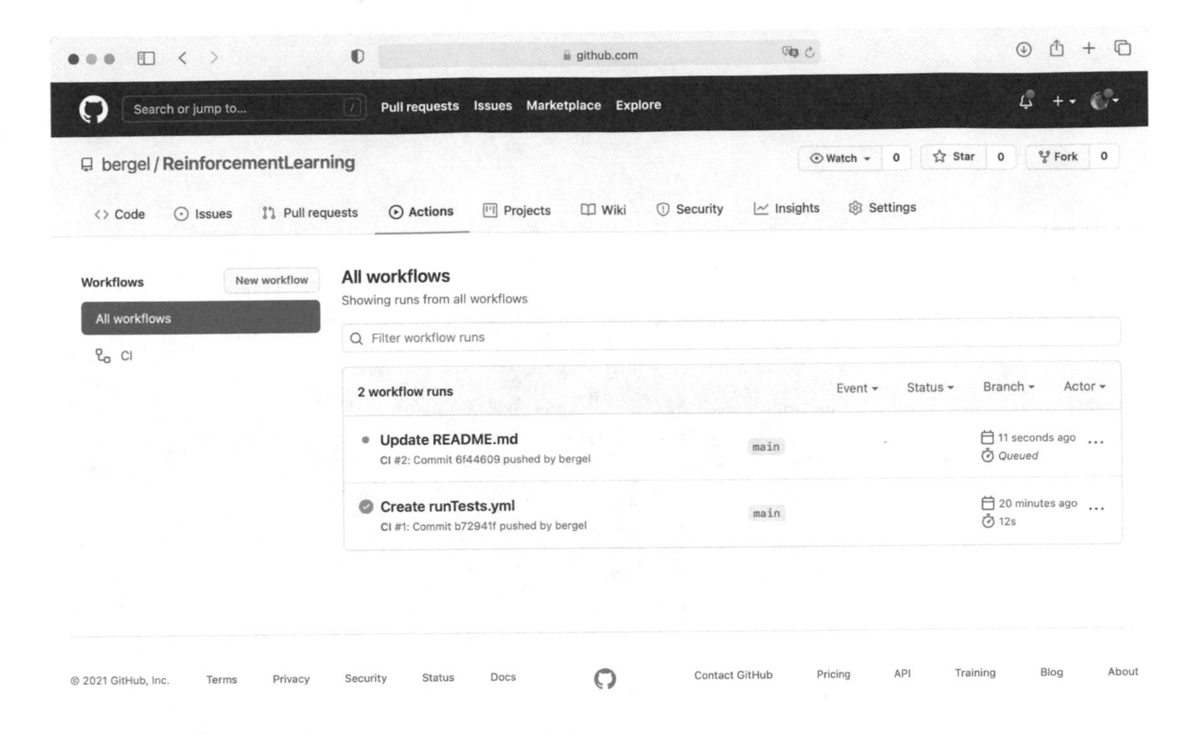

Figure 15-3. *Workflow is executed at each commit*

Figure 15-3 indicates that the workflow was first created when you committed the `runTests.yml` file, and a new workflow runs after editing the `README.md` file.

The large green dot next to the `Create runTests.yml` workflow indicates that the workflow was correctly executed (i.e., it did not produce an error), and the small yellow dot indicates that the workflow triggered by updating the `README.md` file is in the queue and will be executed shortly. By clicking the `Update README.md` workflow name and then clicking the build button, you get the result of the workflow execution (see Figure 15-4).

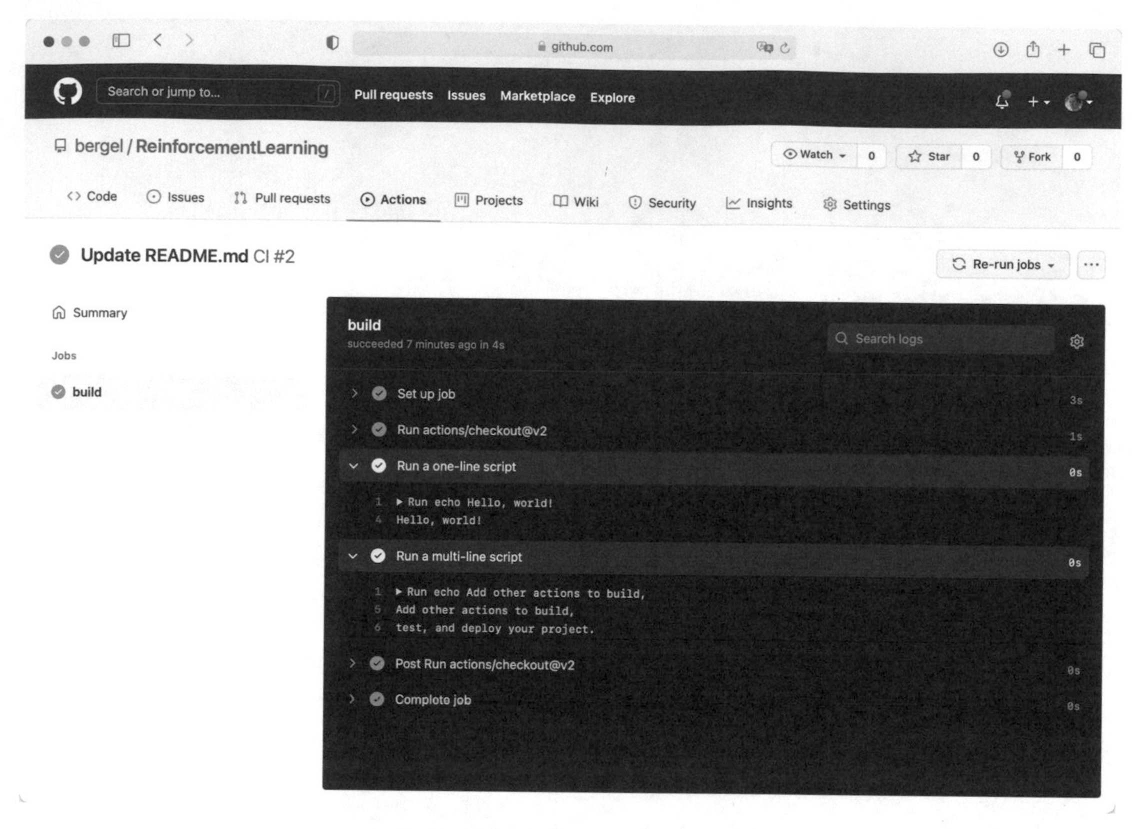

Figure 15-4. *Workflow executing result*

The workflow execution result lists the steps that are executed and their results. You can see that the echo shell scripts commands are executing, as well as their output.

The steps contained in the workflow are extremely simple. At that stage, you may have two questions: Can you replace the echo instruction with arbitrary complex instructions? Where are the echo commands really executed? The advances of cloud computing enable you to use complex instructions in place of the echo instruction, which will be executed on the GitHub cloud. GitHub Actions is a service to run workflows on a cloud, and it comes with a free plan that's currently used it in this chapter.

The next section replaces the echo commands with a significantly more complex set of instructions, in particular, to cover the complete installation and execution of Pharo. Steps contained in a workflow are run on a GitHub-hosted runner. This workflow specifies the host runner to use the latest version of the Ubuntu operating system. The host runner is a two-core CPU, with 7GB of RAM memory and 14GB of SSD disk space. Just what you need to execute a few Roassal scripts.

Running Unit Tests

This workflow simply prints some text, which is not that interesting. In this first example, you will adapt the workflow to run unit tests of a Pharo application. Edit and commit the `runTests.yml` file located in the `.github/workflows` folder as follows (see Figure 15-5):

```
# This is a basic workflow to help you get started with Actions

name: Tests

# Controls when the action will run.
on:
  # Triggers the workflow on push or pull request events but only for the
    main branch
  push:
    branches: [ main ]
  pull_request:
    branches: [ main ]

  # Allows you to run this workflow manually from the Actions tab
  workflow_dispatch:

# A workflow run is made up of one or more jobs that can run sequentially
  or in parallel
jobs:
  # This workflow contains a single job called "build"
  build:
    # The type of runner that the job will run on
    runs-on: ubuntu-latest

    # Steps represent a sequence of tasks that will be executed as part of
      the job
    steps:
      - name: Cancel Previous Runs
        uses: style/cancel-workflow-action@0.9.0
        with:
          access_token: ${{ github.token }}
```

```
      # Checks out your repository under $GITHUB_WORKSPACE, so your job can
        access it
      - uses: actions/checkout@v2

      # Runs a single command using the runners shell
      - name: Run unit tests
        run: bash ./scripts/runTests.sh
```

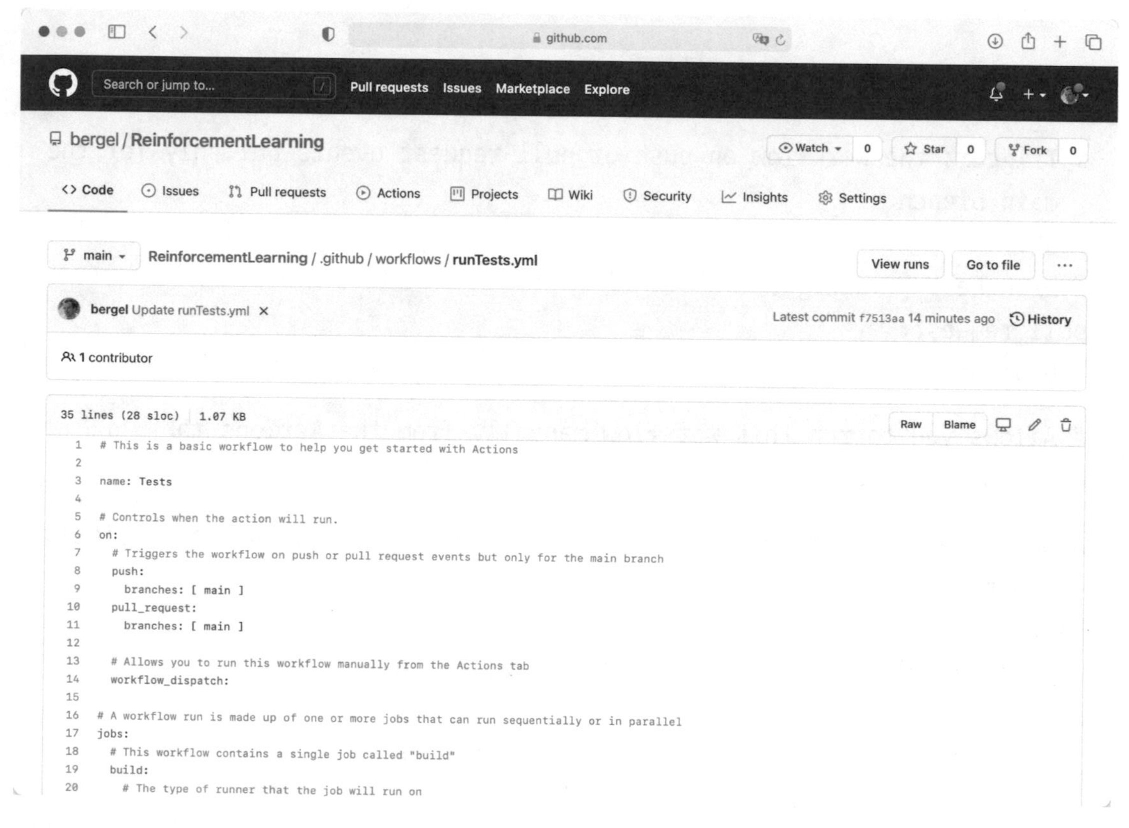

Figure 15-5. *Adapting the workflow*

This new workflow redefines the steps of the workflow. First, the step named `Cancel Previous Runs` prevents a workflow from running when new workflows are queued. This is not an essential step, but it is very handy to avoid executing workflows on old commits. The last line of the script executes the `./scripts/runTests.sh` command. Committing this workflow will produce an error when the workflow is executed because the `./scripts/runTests.sh` script does not exist yet.

You can now create a new folder and a new file called `./scripts/runTests.sh` (i.e., a new folder titled `scripts` is created in the root of the GitHub repository with the `runTests.sh` file in it). It has the following content:

```
curl -L https://get.pharo.org/64/ | bash
./pharo --headless Pharo.image ./scripts/runTests.st

FILE=/tmp/result.txt
if [ ! -f "$FILE" ]; then
    echo "ERROR: $FILE does not exists!"
    exit 1
fi

# Print the result of the tests
cat $FILE

if grep -q ERROR "$FILE"; then
    echo "SOME ERRORS!"
    exit 1
else
    echo "ALL TEST PASSED"
    exit 0
fi
```

The script begins by installing the stable 64-bit version of Pharo. The script then runs Pharo in a headless mode, with no windows or UI, since the GitHub host runner does not have a screen. Use the standard image and provide the Pharo script `./scripts/runTests.st` to run when the image is open. The result of the workflow is based on the exit code returned by the `runTests.sh` file. If the file exits with the value `0`, this means that tests were run and they all passed. If the file exists with the value `1`, an error occurred. The `/tmp/result.txt` file will be produced by `runTests.st` (yes, `runTests.st` and not `runTests.sh`), which is presented in the next section.

Looking at the Actions tab, you can see that the executed workflow produces an error because the`./scripts/runTests.st` file is missing (see Figure 15-6).

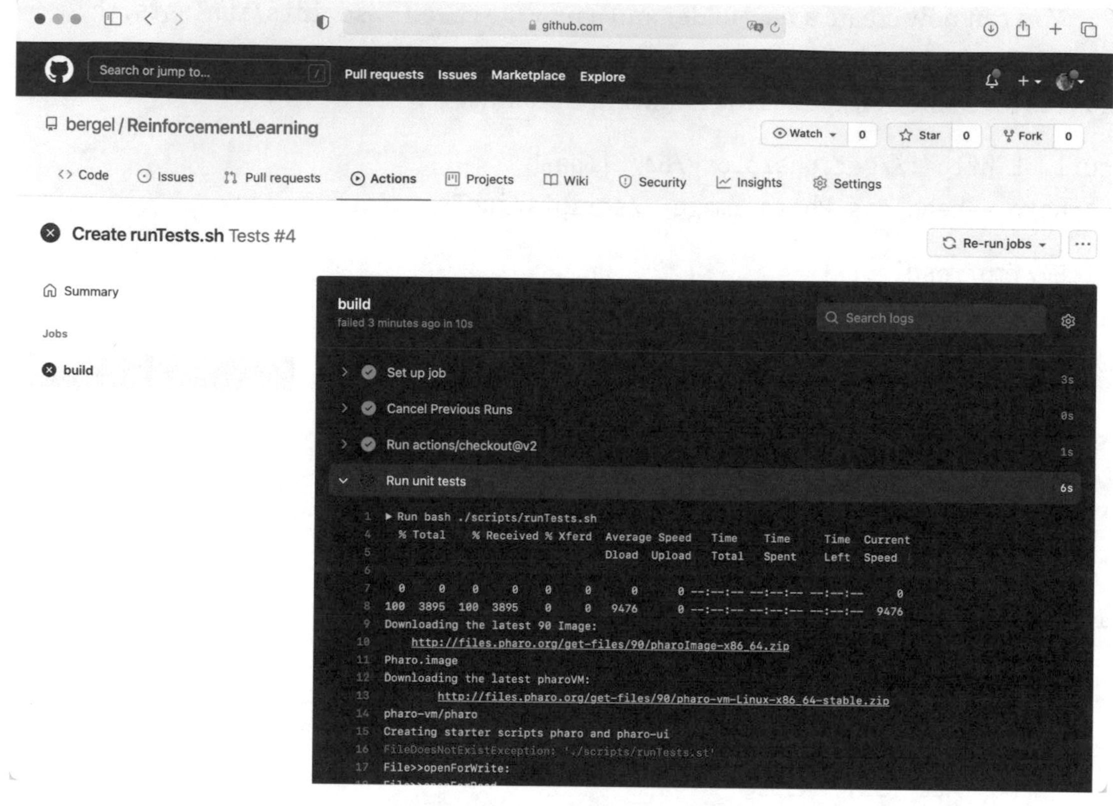

***Figure 15-6.** Error because the workflow is still incomplete*

The next section explains how to complete the workflow and fix this error.

Running Tests

You need to define the `./scripts/runTests.st` file as follows:

```
"We capture all the unit tests in Pharo"
unitTestsBefore := TestCase withAllSubclasses.

"We load the application"
[ Metacello new
    baseline: 'ReinforcementLearning';
    repository: 'github://bergel/ReinforcementLearning:main';
    load ] on: MCMergeOrLoadWarning do: [ :warning | warning load ].
```

```
"We identify the unit tests contained in the loaded application"
unitTests := (TestCase withAllSubclasses copyWithoutAll: unitTestsBefore)
sorted: [ :c1 :c2 | c1 name < c2 name ].

"We create a file that will contain the result of the test execution"
path := '/tmp/result.txt'.
path asFileReference exists ifTrue: [ path asFileReference delete ].
stream := path asFileReference writeStream.

"We run the unit tests"
unitTests do: [ :unitTestClass |
  res := unitTestClass buildSuite run.
  (res hasFailures or: [ res hasErrors ]) ifTrue: [ stream nextPutAll:
  'ERROR: ' ].
  stream nextPutAll: unitTestClass name, ' ', res asString.
  stream crlf
].

"We close the stream and quit Pharo"
stream close.
SmalltalkImage current quitPrimitive
```

The objective of this file is to load the `ReinforcementLearning` application, execute all the tests that belong to it, and produce a file named `/tmp/result.txt` containing the result of the test execution. The Pharo script begins by getting a list of all the unit tests in Pharo. This is used to identify the unit tests that belong to the application to be loaded. The following instruction loads the `ReinforcementLearning` application using Metacello. If you are not aware of what Metacello is or how to write a baseline, I recommend you read the additional online resources given at the end of this chapter.

The `unitTests` variable contains all the unit tests that belong to the `ReinforcementLearning` application. A `/tmp/result.txt` file is then created, which will contain all the results of the test execution. Note that this file is created on the GitHub host runner's virtual computer. The unit tests are then run and the result is written in the temporary file. The script ends by quitting the Pharo image.

At this stage, the workflow should execute properly. This can be verified from the Actions tab (see Figure 15-7).

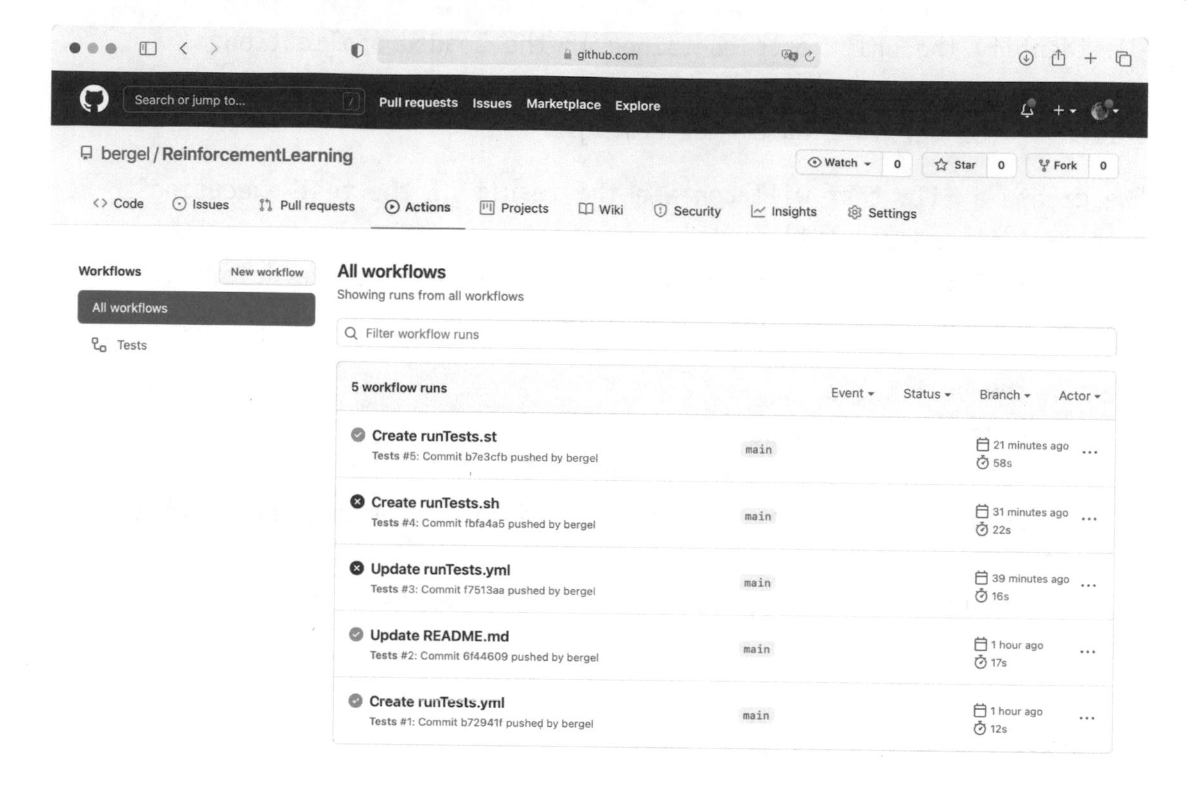

Figure 15-7. *The workflow is now back to green*

Pushing in the repository will run the unit tests that belong to the `ReinforcementLearning` application. The result of the test executions can be seen by clicking the workflow name followed by the Build button (see Figure 15-8).

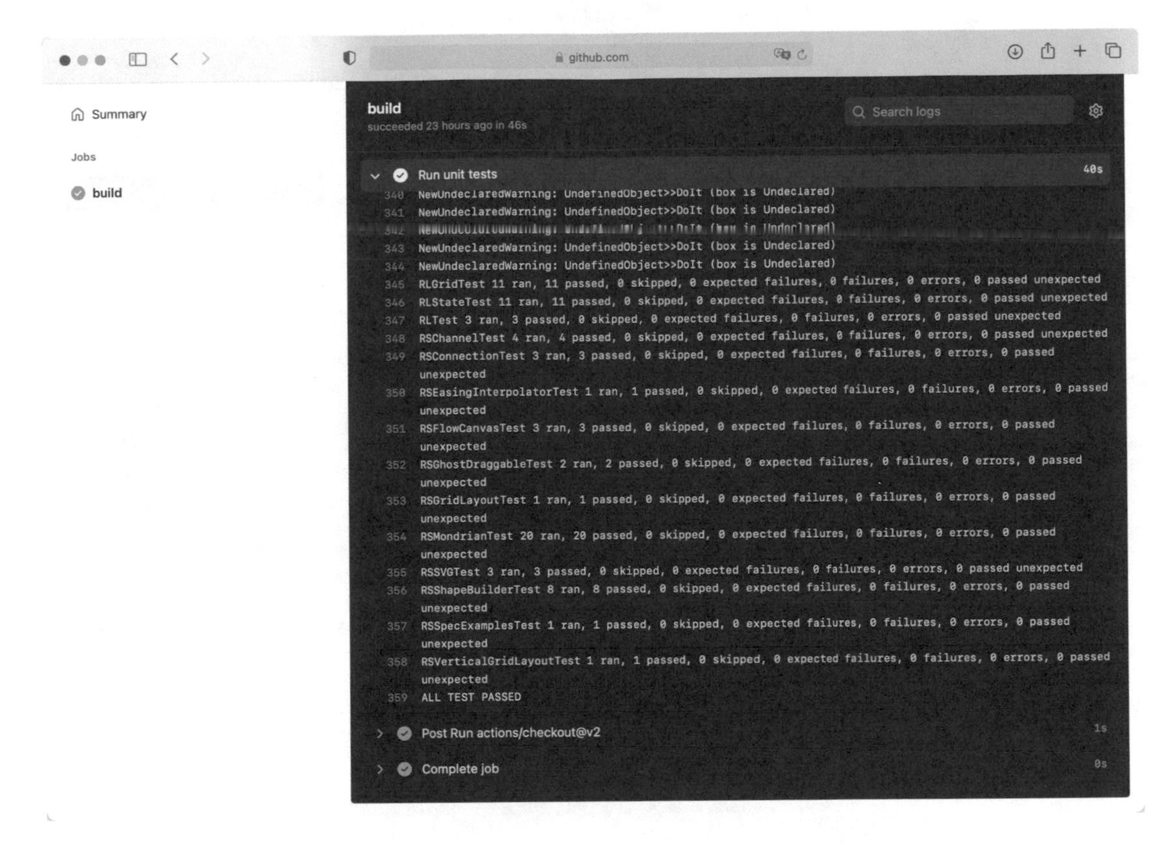

***Figure 15-8.** Verifying the result of the test execution*

A badge is a little icon that can be inserted into the `README.md` file to easily convey a property of the project, such as the result of the test execution. Having a badge that indicates whether the tests successfully executed is a good practice. Selecting a menu of the workflow shows a markdown instruction that displays a badge (see Figure 15-9).

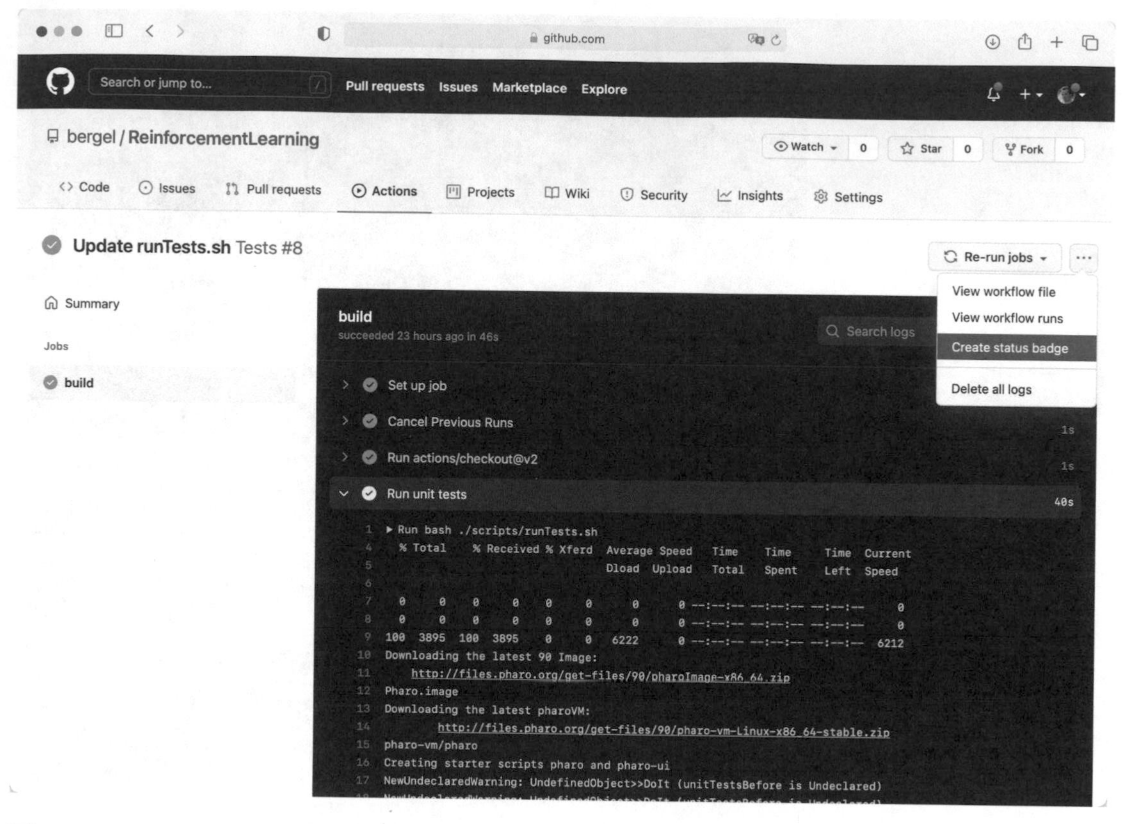

Figure 15-9. *Obtaining a badge*

The markdown instruction that has to be inserted at the beginning of the `README.md` file should look like this:

```
[![Tests](https://github.com/bergel/ReinforcementLearning/actions/
workflows/runTests.yml/badge.svg)](https://github.com/bergel/
ReinforcementLearning/actions/workflows/runTests.yml)
```

This instruction just needs to be added to the `README.md` file. It will show a badge reflecting the success of the unit tests execution (see Figure 15-10).

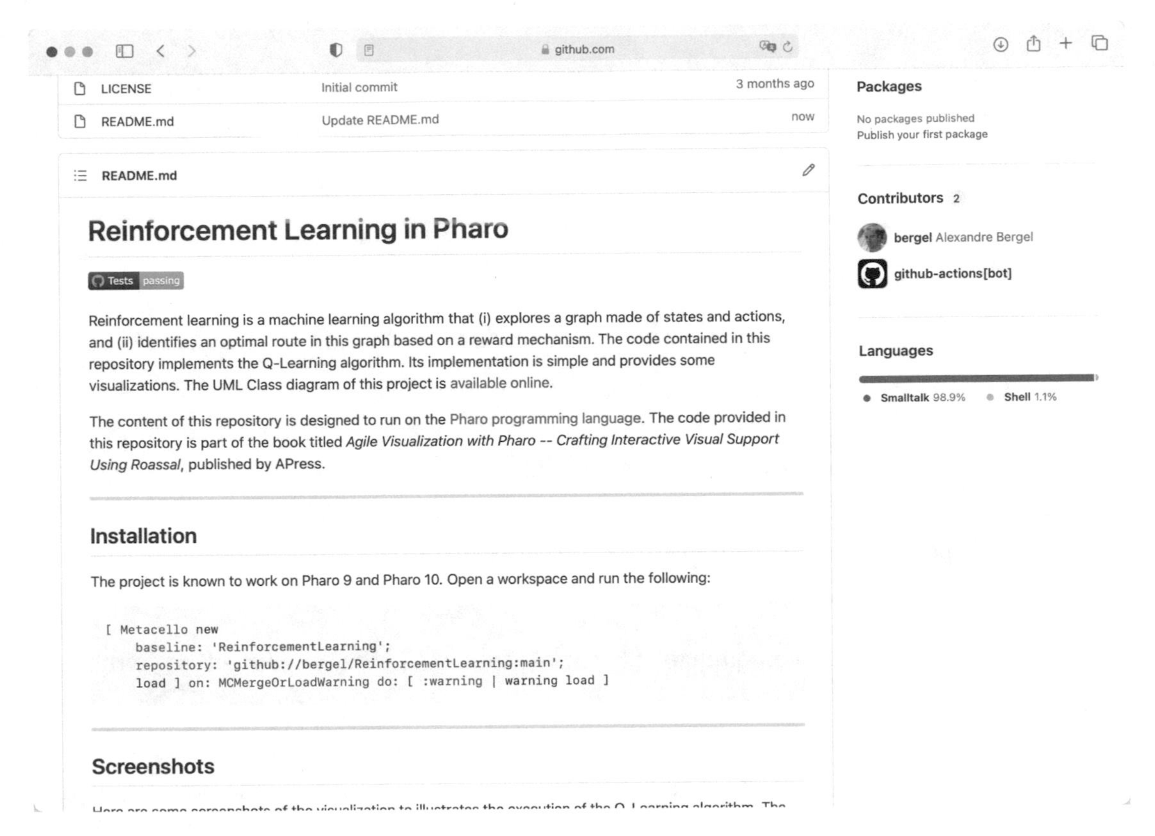

Figure 15-10. *The README.md file and badge*

Visualizing the UML Class Diagram

As a first visualization, you will generate a class diagram of the classes contained in the repository. Adding this visualization simply consists of redoing what you previously did with new scripts. In particular, you define a new workflow titled `.github/workflows/visualizeClassDiagram.yml`:

```
# This is a basic workflow to help you get started with Actions

name: UML Class diagram

# Controls when the action will run.
on:
  # Triggers the workflow on push or pull request events but only for the
    master branch
```

```
  push:
    branches: [ main ]
  pull_request:
    branches: [ main ]

  # Allows you to run this workflow manually from the Actions tab
  workflow_dispatch:

# A workflow run is made up of one or more jobs that can run sequentially
  or in parallel
jobs:
  # This workflow contains a single job called "build"
  build:
    # The type of runner that the job will run on
    runs-on: ubuntu-latest

    # Steps represent a sequence of tasks that will be executed as part of
      the job
    steps:
      # Checks out your repository under $GITHUB_WORKSPACE, so your job can
        access it
      - uses: actions/checkout@v2
        with:
          persist-credentials: false # otherwise, the token used is the
          GITHUB_TOKEN, instead of your personal token
          fetch-depth: 0 # otherwise, you will fail to push refs to dest
          repo

      - name: Producing class diagram
        run: bash ./scripts/visualizeClassDiagram.sh
      - name: Push the picture
        run: |
          mkdir -p ci_data
          cd ci_data
          mv ../uml.png .
          git add uml.png
      - name: Commit files
```

```
      run: |
        git config --local user.email "41898282+github-actions[bot]@
        users.noreply.github.com"
        git config --local user.name "github-actions[bot]"
        git diff-index --quiet HEAD || git commit -m "Add changes" -a
    - name: Push changes
      uses: ad-m/github-push-action@master
      with:
        github_token: ${{ secrets.GITHUB_TOKEN }}
        branch: ${{ github.ref }}
```

This workflow definition is significantly longer than the previous workflow you built. The step named `Producing class diagram` simply runs the `./scripts/visualizeClassDiagram.sh` script, which will create a `uml.png` file. However, several steps are new:

- The `Push the picture` step first creates a folder called `ci_data` at the root of the GitHub repository. This folder will keep all the relevant data, such as a picture of the class diagram. The `uml.png` file is moved in `ci_data` and then added to the Git repository.
- The `Commit files` step commits what has been added to the Git repository.
- The `Push changes` step pushes to the GitHub repository what has been committed.

The complexity of this script comes from a number of points:

- The commits may fail if the `uml.png` file is exactly the same as the version in the repository. By default, Git does not allow empty commits or committing exactly the same file. The `git diff-index --quiet HEAD` instruction is a (non-obvious) way to force the workflow not to fail if the commit cannot be done.
- The commits of the `uml.png` file are authored by a bot of GitHub Actions, and not by the author of the GitHub repository. Before committing, the proper credentials need to be set using `git config`.

The ./scripts/visualizeClassDiagram.sh file is defined as follows:

```
curl -L https://get.pharo.org/64/alpha+vm | bash
./pharo --headless Pharo.image ./scripts/visualizeClassDiagram.st
```

The ./scripts/visualizeClassDiagram.st file is defined as follows:

```
[ Metacello new
    baseline: 'Roassal3';
    repository: 'github://ObjectProfile/Roassal3';
    load: 'Full' ] on: MCMergeOrLoadWarning do: [:warning | warning load ].

[ Metacello new
    baseline: 'Roassal3Exporters';
    repository: 'github://ObjectProfile/Roassal3Exporters';
    load ] on: MCMergeOrLoadWarning do: [:warning | warning load ].

[ Metacello new
    baseline: 'ReinforcementLearning';
    repository: 'github://bergel/ReinforcementLearning:main';
    load ] on: MCMergeOrLoadWarning do: [:warning | warning load ].

packageName := 'ReinforcementLearning'.
uml := RSUMLClassBuilder new.
uml classes: (RPackageOrganizer default packageNamed: packageName)
definedClasses.
uml build.

uml canvas extent: 1000 @ 1000.
RSPNGExporter new
  canvas: uml canvas;
  exportToFile: 'uml.png' asFileReference.

SmalltalkImage current quitPrimitive
```

It simply loads the full version of Roassal, the exporters, and the ReinforcementLearning project. It then uses the RSUMLClassBuilder class to create a class diagram, which is stored in the ./ci_data/uml.png file. You can produce a badge to indicate that the class diagram was correctly built and to give easy access to it.

You can use the markdown instruction as a badge:

```
[![UML Class diagram](https://github.com/bergel/ReinforcementLearning/
actions/workflows/visualizeClassDiagram.yml/badge.svg)](https://github.com/
bergel/ReinforcementLearning/blob/main/ci_data/uml.png)
```

Naturally, you should substitute the link to your own repository's `uml.png`. This instruction is a slight adaption of the instruction produced by GitHub. This version simply points to the picture itself, instead of pointing to the workflow result. The `README.md`, containing the two badges, should look like Figure 15-11.

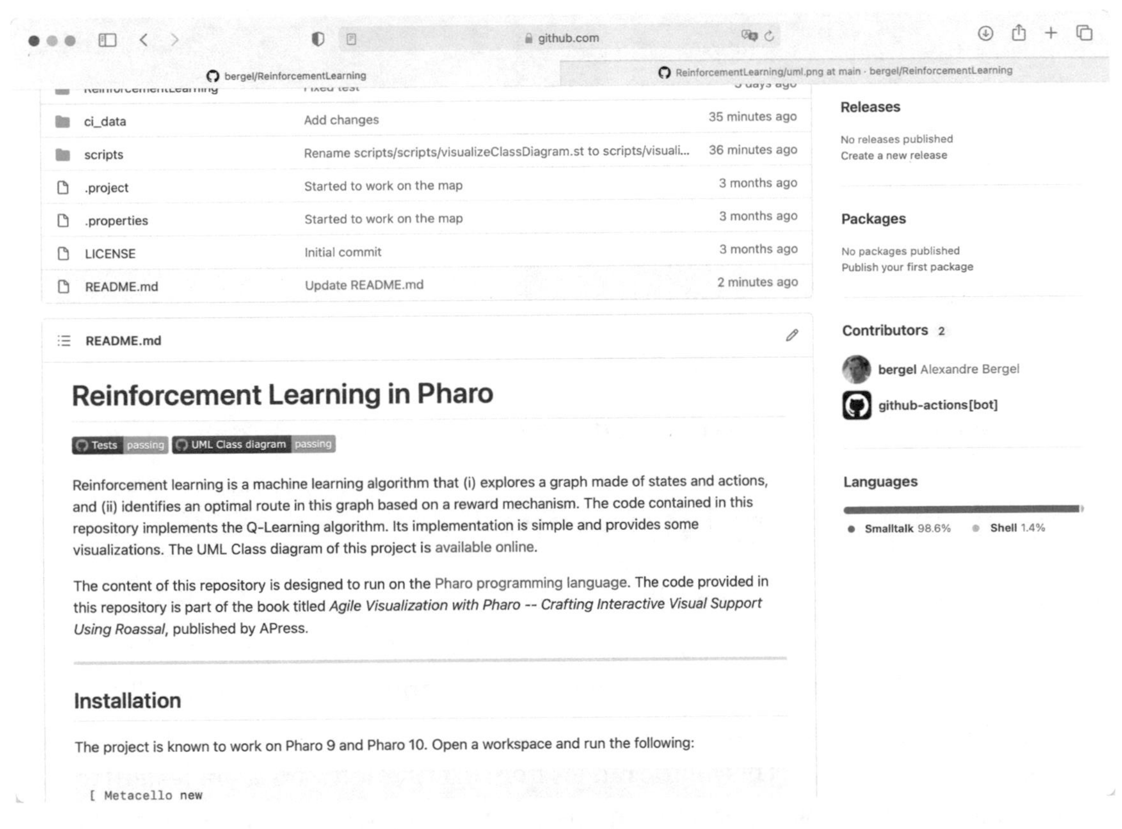

Figure 15-11. *README.md and its two badges*

Clicking the badge should lead you to the UML class diagram itself (see Figure 15-12).

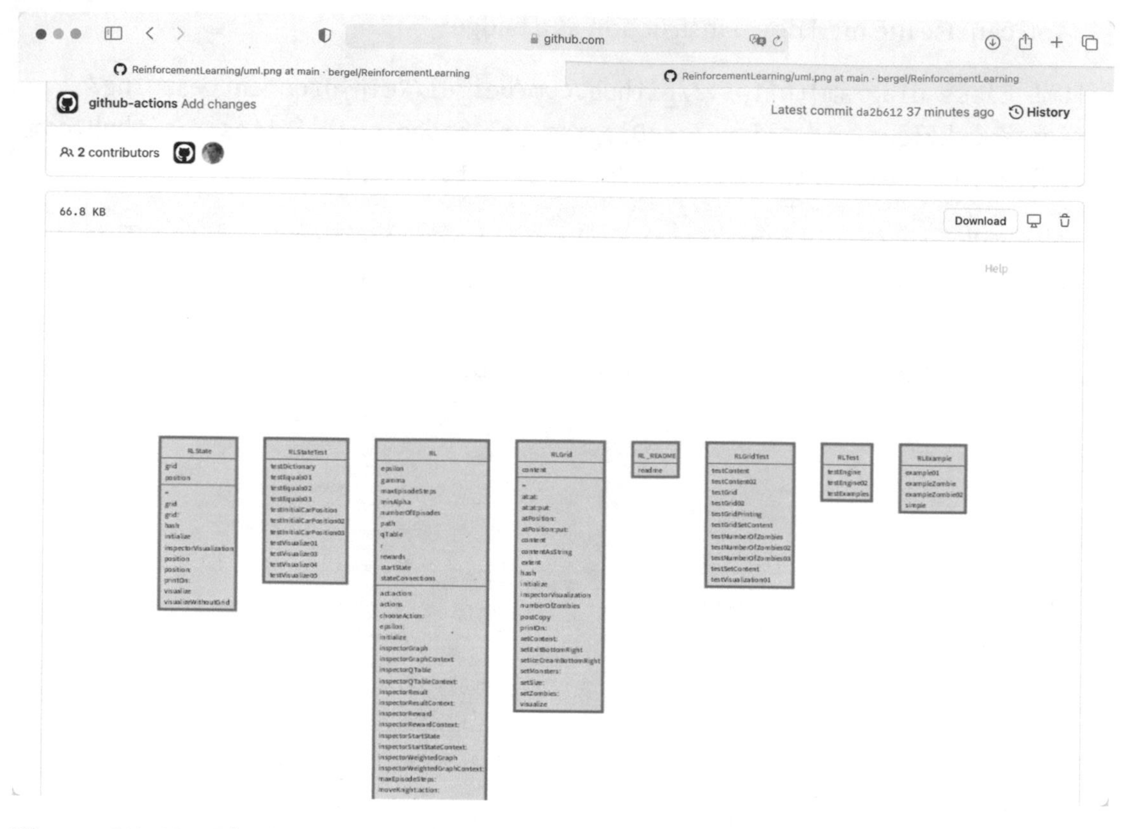

Figure 15-12. *The UML class diagram of the ReinforcementLearning project*

Visualizing the Test Coverage

You have seen two uses of GitHub Actions so far. First, you defined a workflow that simply runs the tests. Second, you produced a UML class diagram. In this section, you will combine these two experiences by visually representing the test coverage. The test coverage is a classical metric of a software system that represents the portion of source code covered by unit tests. This metric ranges from 0.0 (the software is not tested) to 1.0 (the software is completely tested). In this example, you will see how to provide this metric and show which part of the code is not tested. Similarly, you define three files:

- `.github/workflows/visualizeCoverage.yml` describes the GitHub Actions workflow.
- `./scripts/visualizeCoverage.sh` launches Pharo.
- `./scripts/visualizeCoverage.st` runs the unit tests, computes the test coverage, and generates the visualization.

Define the `.github/workflows/visualizeCoverage.yml` file as follows:

```
# This is a basic workflow to help you get started with Actions

name: Visualize coverage

# Controls when the action will run.
on:
  # Triggers the workflow on push or pull request events but only for the
    master branch
  push:
    branches: [ main ]
  pull_request:
    branches: [ main ]

  # Allows you to run this workflow manually from the Actions tab
  workflow_dispatch:

# A workflow run is made up of one or more jobs that can run sequentially
  or in parallel
jobs:
  # This workflow contains a single job called "build"
  build:
    # The type of runner that the job will run on
    runs-on: ubuntu-latest

    # Steps represent a sequence of tasks that will be executed as part of
      the job
    steps:
      # Checks out your repository under $GITHUB_WORKSPACE, so your job can
        access it
      - uses: actions/checkout@v2
        with:
          persist-credentials: false # otherwise, the token used is the
          GITHUB_TOKEN, instead of your personal token
          fetch-depth: 0 # otherwise, you will fail to push refs to dest
          repo

      # Runs a single command using the runners shell
```

```
      - name: Run unit tests
        run: bash ./scripts/visualizeCoverage.sh

      - name: Push the picture
        run: |
          mkdir -p ci_data
          cd ci_data
          mv ../coverage.png .
          git add coverage.png
          cat /tmp/ratio
          wget "https://img.shields.io/badge/Coverage-`cat /tmp/ratio`-
          green"
          mv "Coverage-`cat /tmp/ratio`-green" coverageBadge.svg
          git add coverageBadge.svg
      - name: Commit files
        run: |
          git config --local user.email "41898282+github-actions[bot]
          @users.noreply.github.com"
          git config --local user.name "github-actions[bot]"
          git diff-index --quiet HEAD || git commit -m "Add changes" -a
          git pull
      - name: Push changes
        uses: ad-m/github-push-action@master
        with:
          github_token: ${{ secrets.GITHUB_TOKEN }}
          branch: ${{ github.ref }}
```

The step named `Run unit tests` launches Pharo. As described next, the step produces a picture named `coverage.png` and a `/tmp/ratio` file containing the test coverage ratio. The `Push in the repository` step adds the `./ci_data/coverage.png` and `./ci_data/coverageBadge.` Svg files. This SVG file represents a customized badge that will be shown in the `README.md` file. The badge is created by an online service called `https://shields.io.` For example, if the test coverage is 93.45%, the badge that is obtained from `https://shields.io` is `https://img.shields.io/badge/Coverage-93.45-green.`

The ./scripts/visualizeCoverage.sh file launches Pharo with the instruction to run the unit tests, compute the coverage, and produce the necessary files:

```
curl -L https://get.pharo.org/64/alpha+vm | bash
./pharo --headless Pharo.image ./scripts/visualizeCoverage.st
```

The ./scripts/visualizeCoverage.st file is defined as follows:

```
"Loading all the tool suite"
[ Metacello new
  baseline: 'Spy2';
  repository: 'github://ObjectProfile/Spy2';
  load: 'HapaoCore' ] on: MCMergeOrLoadWarning do: [:warning | warning
  load ].

[ Metacello new
    baseline: 'Roassal3';
    repository: 'github://ObjectProfile/Roassal3';
    load: 'Full' ] on: MCMergeOrLoadWarning do: [:warning | warning load ].

[ Metacello new
    baseline: 'Roassal3Exporters';
    repository: 'github://ObjectProfile/Roassal3Exporters';
    load ] on: MCMergeOrLoadWarning do: [:warning | warning load ].

[ Metacello new
    baseline: 'ReinforcementLearning';
    repository: 'github://bergel/ReinforcementLearning:main';
    load ] on: MCMergeOrLoadWarning do: [:warning | warning load ].

"Configuring the visualization"
shouldClassHaveName := true.
classNameHeight := 3.
packageNameAsRegExp := 'ReinforcementLearning*'.
numberOfMethodsToList := 10.

"Script that visualize the coverage"
profiler := 'Hapao2' asClass runTestsForPackagesMatching:
packageNameAsRegExp.
```

```
classSizeScale := 'NSScale' asClass linear range: #(5 30); domain: { 0 .
(profiler allClasses max: #numberOfMethods) }.
classColorCoverage := 'NSScale' asClass linear range: {Color red . Color
black}; domain: { 0 . 100 }.
m := 'RSMondrian' asClass new.
m shape labeled: #packageName; color: Color white; borderColor: Color gray.
m nodes: profiler packages forEach: [ :pak |
  m shape box
    size: [ :cls | classSizeScale scale: cls numberOfMethods ];
    color: [ :cls | classColorCoverage scale: cls coverage ];
    if: #isTestClass color: Color green darker.

  someBoxes := m nodes: pak classes.
  shouldClassHaveName ifTrue: [ someBoxes @ (RSLabeled new fontSize:
  classNameHeight) ].
  m orthoVerticalLine connectFrom: #superclass.
  m layout tidyTree
].
m build.

"Indicate the overall coverage"
lbl := RSLabel text: 'Ratio of covered methods = ', (profiler coverage
asString), ' %'.
RSLocation move: lbl above: m canvas nodes.
m canvas add: lbl.

"Show a few uncovered methods"
uncoveredMethods := profiler noncoveredMethods.
uncoveredMethods notEmpty ifTrue: [
  labels := RSLabel models: (uncoveredMethods copyFrom: 1 to:
  (uncoveredMethods size min: numberOfMethodsToList)).
  RSCellLayout new lineItemsCount: 2; on: labels.
  RSLocation move: labels below: m canvas nodes.
  labels translateBy: 0 @ 40.
  labels @ RSHighlightable red.
  m canvas addAll: labels.
```

```
    titleLabel := RSLabel new text: 'Some uncovered methods:'; color: Color
    black.
    m canvas add: titleLabel.
    RSLocation move: titleLabel above: labels.
].

"Exporting the picture"
m canvas extent: 1000 @ 1000.
RSPNGExporter new
    canvas: m canvas;
    exportToFile: 'coverage.png' asFileReference.

"We create a file that will contain the result of the test coverage"
path := '/tmp/ratio'.
path asFileReference exists ifTrue: [ path asFileReference delete ].
stream := path asFileReference writeStream.
stream nextPutAll: profiler coverage asString.
stream close.

"Quitting Pharo"
SmalltalkImage current quitPrimitive
```

Figure 15-13 shows an example of the visualization that is built by this script. Each package of the analyzed application is represented as a large box, with its name at the top. The overall test coverage ratio is indicated at the very top. Each class defined in the package is represented as a box. The size of the box indicates the number of methods it contains. For example, the `RL` class is the largest class and contains the most methods. Green classes are unit tests. The color of a non-unit test class ranges from black to red. Black indicates that the class is well tested, and red indicates that it is not tested. Red classes should therefore be the focus of further testing efforts.

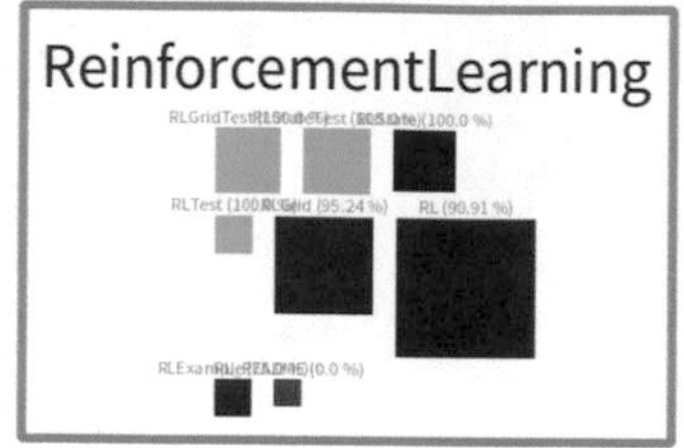

Figure 15-13. *Visualizing the test coverage*

The file loads the necessary tools. In particular, the test coverage is measured using the Spy and Hapao tools. The script then defines four variables:

- `shouldClassHaveName` indicates whether the class names should have names. If the analyzed application is large, the names may unnecessarily clutter the visualization.
- `classNameHeight` sets the height of the font size of the class name. If the font is too large, the class names may overlap.
- `packageNameAsRegExp` is a regular expression that indicates the packages that are analyzed by Spy and Hapao. The `'ReinforcementLearning*'` value indicates that all the packages begin with `ReinforcementLearning`. Only one package is matched in this case.
- `numberOfMethodsToList` is the number of untested methods to list. This list helps increase the test coverage by listing some methods that are left untested.

The `'Hapao2'asClass runTestsForPackagesMatching: packageNameAsRegExp` expression uses the Hapao component of the Spy project to run the tests and measure the test coverage. `RSMondrian` is then used to build the visualization.

You can now add a customized badge to the front page of the project. Add the following instruction to the beginning of the README.md file:

```
[![Coverage](https://raw.githubusercontent.com/bergel/
ReinforcementLearning/main/ci_data/coverageBadge.svg)](https://github.com/
bergel/ReinforcementLearning/blob/main/ci_data/coverage.png)
```

You have now completed the generation of the test coverage visualization.

What Have You Learned in This Chapter?

This chapter explored the fascinating topic of continuous integration and delivery. It illustrated how Roassal helps produce visual reports, produced at each change in the repository. In particular, the chapter provided:

- A minimal description of GitHub Actions.
- A simple scenario on how to create a workflow.
- Two non-trivial visualizations that indicate relevant aspects of a software system written in Pharo.

A number of online resources are available if you want to learn more about GitHub Actions. In particular, `https://docs.github.com/en/actions/learn-github-actions` is a good starting point.

This chapter assumes that you are familiar with numerous topics, including baselines and Metacello. Metacello and baselines are explained in the book *Deep Into Pharo* (see `http://deepintopharo.com`).

Index

A. Bergel, *Agile Visualization with Pharo*, https://doi.org/10.1007/978-1-4842-7161-2

S

T, U

V, W, X, Y, Z

Printed in the United States
by Baker & Taylor Publisher Services